HUMAN DEVELOPMENT
AND LEARNING

HUMAN DEVELOPMENT AND LEARNING

by

Robert D. Strom
Harold W. Bernard
Shirley K. Strom

Arizona State University

BF
713
.S77
1987
WES7

HUMAN SCIENCES PRESS, INC.
72 FIFTH AVENUE
NEW YORK, N.Y. 10011-8004   (212) 243-6000

ASU WEST LIBRARY

Copyright © 1987 by Human Sciences Press, Inc.

72 Fifth Avenue, New York, New York 10011

All rights reserved. No part of this work may be reproduced or utilized in any form or by any means, electronic or mechanical, including photocopying, microfilm and recording, or by any information storage and retrieval system without permission in writing from the publisher.

Printed in the United States of America
987654321

Library of Congress Cataloging-in-Publication Data

Strom, Robert D.
  Human development and learning.

  Bibliography: p.
  Includes index.
  1. Developmental psychology. 2. Learning.
I. Bernard, Harold W. (Harold Wright), 1908–
II. Strom, Shirley K. III. Title. [DNLM: 1. Human Development. 2. Learning. BF 713 S921h]
BF713.S77 1987   155   86-21393
ISBN 0-89885-360-5

CONTENTS

CHAPTER 1 UNDERSTANDING THE LIFESPAN	**17**
SOURCES OF HUMAN DEVELOPMENT DATA	18
Forerunners of Current Knowledge	18
Biographical Studies	18
Questionnaire Studies	18
Cross-Sectional Studies	19
Longitudinal Investigations	19
Experimental Methods	20
Clinical Studies	20
Using Data on Development	21
VIEWPOINTS ON DEVELOPMENT	21
Hereditarian Perspective	21
Behavioristic Perspective	23
Gestalt-Field Orientation	24
Psychoanalytic Orientations	26
Humanistic Psychology	27
Social Learning Theory	29
Information Processing	30
DEVELOPMENTAL AGE STAGES	31
Ancestry	31
Infancy—Birth to Age 2	32

6   HUMAN DEVELOPMENT AND LEARNING

Early Childhood—Ages 2 to 6	34
Middle Childhood—Ages 6 to 9	35
Later Childhood—Ages 9 to 12	36
Adolescence—Ages 12 to 20	37
Early Adulthood—Ages 20 to 40	39
Middle Adulthood—Ages 40 to 60	40
Later Adulthood—Ages 60 to Death	41
PRINCIPLES OF DEVELOPMENT	43
Development Is Interactive	43
Multiple Causation	43
Initial Rapidity	43
Differentiation and Integration	44
Drive for Homeostatic Balance	44
Drive for Heterostatic Activity	45
Development is Continuous	46
Training Influences Development	46
Development is Sequential	46
Correlation and Compensation	47
CHAPTER 2 MAINTAINING HEALTH AND WELLNESS	**49**
WELLNESS: AN ISSUE OF PERVASIVE RELEVANCE	49
PRENATAL DEVELOPMENT	52
Critical Periods	52
Avoidable Prenatal Hazards	53
Prenatal Diagnoses	54
BIRTH	55
Delivery Preparedness	55
Birth Trauma and Mortality	56
INFANCY	57
Assessment of Growth	57
Hunger and Nutrition	57
EARLY CHILDHOOD	58
Children's Needs	59
Child Abuse	60
MIDDLE CHILDHOOD	61
Growth and Health	61
Physical Fitness	61
Obesity	62
LATER CHILDHOOD	63
Physical Growth	63

Vision and Hearing	64
ADOLESCENCE	65
Physical Characteristics of Adolescents	65
Adolescent Glandular Development	66
Accidents and Suicides	66
EARLY ADULTHOOD	68
Physical Characteristics	68
Anorexia and Bulimia	70
Diet and Disease	71
Sexually Transmitted Diseases	72
MIDDLE ADULTHOOD	73
Functional Physical Efficiency	73
Sensory Acuity	74
Sexual Function and Behavior	74
LATER ADULTHOOD	76
Potential Lifespan	76
Physical Changes in Aging	77
CHAPTER 3 DEVELOPING INTELLECTUAL ABILITIES	**79**
INTELLIGENCE AND THINKING	80
Changing Views of Intelligence	80
Implications of Varied Intelligences	81
Brain Hemispheres and Learning	82
Age-Stages in Cognitive Development	83
INFANCY	84
Nutrition and Brain Development	84
Sensorimotor Thinking	86
EARLY CHILDHOOD	86
Brain Growth Spurts	86
Preoperational Thinking	88
MIDDLE AND LATER CHILDHOOD	89
Concrete Operational Thinking	89
Styles of Learning	90
Creativity and Intelligence	92
ADOLESCENCE	93
Formal Operational Thinking	93
Teaching Implications of Piaget's Theory	93
Emergence of Cognitive Subabilities	94
Brain Reorganization	95
Assessment of Intelligence	96

8 HUMAN DEVELOPMENT AND LEARNING

Cessation of Cognitive Development	97
Identifying the Gifted and Talented	98
EARLY AND MIDDLE ADULTHOOD	101
Creativity and Job Success	101
The Creative Thinking Process	102
LATER ADULTHOOD	105
Mental Functioning and Senility	105
Intellectual Performance and Aging	107
CHAPTER 4 SUPPORTING EMOTIONAL GROWTH	**109**
THEORIES OF EMOTION	110
James-Lange Theory	110
Thalamic Theory	110
Activity Theory	111
Cognitive Aspects of Emotion	112
INFANCY	112
Developing Positive Emotions	112
Bonding and Attachment	113
Lifelong Attitudinal Crises	113
Trust versus Mistrust	114
EARLY CHILDHOOD	115
Autonomy versus Shame and Doubt	115
Handling Anger and Frustration	116
Initiative versus Guilt: Feelings about School	117
MIDDLE AND LATER CHILDHOOD	118
Industry versus Inferiority	118
Defining Success and Failure	118
Overcomers	120
Learning to Discuss Feelings	120
Sharing Fears	121
Drugs and Anxiety	123
ADOLESCENCE	124
Identity versus Role Confusion	124
Maintaining Self-Concept: The Role of Defense Mechanisms	125
Peer Relations and Locus of Control	127
EARLY ADULTHOOD	128
Theories of Adult Emotion	128
Intimacy versus Isolation	129
Enjoying Parenting	130

MIDDLE AGE	132
Generativity versus Stagnation	132
Longlasting Marriages	133
Surmounting Loneliness	136
LATER ADULTHOOD	136
Contrasts Characterize Later Life	136
Ego Integrity versus Despair	137
Attitudes Toward Death and Dying	138
CHAPTER 5 PURSUING LIFELONG LEARNING	**140**
Learning in a Past-Oriented Society	140
Learning in a Present-Oriented Society	141
Learning in a Future-Oriented Society	142
Schooling in the Past	143
Schooling and the Present	143
Schooling and the Future	144
Learning to Reintegrate Society	145
INFANCY	146
The Infant Intellect	146
Learning Through Play	147
EARLY AND MIDDLE CHILDHOOD	148
Creativity and Solitude	148
Learning to be Alone	149
Education and Creativity	151
LATER CHILDHOOD	152
Peer Teaching and Learning	152
Improving Homework Practices	154
Learning about Aging	156
ADOLESCENCE	157
Instrumental and Expressive Curriculum	157
School Success and Failure	158
A Broader View of Basic Curriculum	159
EARLY AND MIDDLE ADULTHOOD	159
Higher Education in Adulthood	159
The Corporate Classroom	161
Readiness in Conflict	163
Changing Views of Conflict	163
LATER ADULTHOOD	166
Access to Learning	166
Basic Curriculum for the Elderly	168

CHAPTER 6   FACILITATING POSITIVE MOTIVATION   170

MEANINGS OF MOTIVATION   171
    Multiple Explanations of Motivation   171
    Hierarchy of Needs   171
INFANCY   173
    Biology of Motivation   173
    Attachment   173
    Reinforcement   174
EARLY CHILDHOOD   175
    Fantasy Play and Values   175
MIDDLE CHILDHOOD   177
    The Pygmalion Effect   177
    Motivation in School   178
    Peer Approval   180
    Parental Pressure   180
LATER CHILDHOOD   181
    Peers, Parents and Planning   181
    Guidance and Goal-setting as Motivators   182
ADOLESCENCE   183
    Boredom-Responsibility   183
    Giftedness and Motivation   185
    Handicaps and Motivation   185
    Motivation Based on Adolescent Needs   187
EARLY ADULTHOOD   188
    New Definitions of Success   188
    Sexuality and Communication   188
    Success, Intimacy and a New Work Ethic   189
    Avoiding Narcissism   191
MIDDLE ADULTHOOD   191
    The Empty-Nest Syndrome   191
    And the Not-So-Empty Nest   192
    Menopause and Climacteric   193
LATER ADULTHOOD   195
    Imaging and Self Assessment   196
    Isolation, Communication and Motivation   196
    Grandparenting   197

CHAPTER 7   COPING WITH STRESS   199

THE CONCEPT OF STRESS   200
    Meanings of Stress   200

Varied Vulnerability	201
Physiology and Drugs	202
Coping and Stress Reduction	203
INFANCY	203
The Needs of Infants	203
Symptoms of Stress	205
Threat to Attachment	205
EARLY CHILDHOOD	206
Aggression and Withdrawal	206
Ethnicity and Stress	207
Intimidation by Peers	208
MIDDLE CHILDHOOD	209
Divorce and Separation	209
Parent Cooperation	210
School Support	210
LATER CHILDHOOD	212
Changing Nature of Stress	212
Meditation and Relaxation	212
Transescence	213
ADOLESCENCE	214
Emphasis on Coping Skills	214
Adolescent Wisdom	215
Boredom and Purpose	217
EARLY ADULTHOOD	218
Adulthood: A Vaguely Defined Stage	218
Multiple Challenges	218
The Personal Role in Early Adult Stress	219
Occupational and Domestic Burnout	220
Choosing a Lifestyle	221
MIDDLE AGE	222
The Need for Self Evaluation	222
Focus on Personal Development	224
LATER ADULTHOOD	225
Sources of Stress	225
Perspectives on Stress	227
CHAPTER 8 ACQUIRING SOCIAL COMPETENCE	**229**
SOCIALIZATION IN PERSPECTIVE	230
A Concept of Socialization	230
INFANCY	231
Social awareness	231

Attachment and Trust	232
EARLY CHILDHOOD	233
Territoriality and Early Socialization	233
Dominion Play and Peer Relationships	234
Guidelines for Dominion Play	236
Social Incompetence and Misbehavior	237
MIDDLE AND LATER CHILDHOOD	237
Belonging and Social Prejudice	237
Accepting the Handicapped	239
Values Clarification	241
Latchkey Children	243
ADOLESCENCE	245
Self Discipline and Social Responsibility	245
Self Evaluation and Self Control	246
The Just Community	248
EARLY ADULTHOOD	249
Developmental Challenges	249
Self Concept and Identity	250
MIDDLE AGE	252
Emergent Perspectives: The Need for Choice	252
Caring for Aging Parents	253
LATER ADULTHOOD	255
Peers and Retirement Communities	255
The Grandparent Role	257
CHAPTER 9 FOSTERING EFFECTIVE COMMUNICATION	**259**
THE MEANING AND ROLE OF COMMUNICATION	260
The Concept of Communication	260
The Vital Importance of Communication	261
Levels of Discourse	263
INFANCY	265
Amazing Skills of Infants	265
Language Development in Action	266
EARLY CHILDHOOD	268
Family Televiewing	268
MIDDLE AND LATER CHILDHOOD	270
School Life and Communication	270
Learning to Read	273
ADOLESCENCE	274
Changing Roles for Parents and Adolescents	274

Steps Toward Improved Communication	275
Levels of Communicative Dialogue	277
Characteristics of Communication	278
EARLY AND MIDDLE ADULTHOOD	279
Communication, Friendship and Therapy	279
Marriage and Communication	280
The Art of Listening	282
Developmental Groups	283
LATER ADULTHOOD	284
Maintaining and Reforming Interpersonal Contacts	284
Family Ties and Individual Perspective	285
Arranging Intergenerational Communication	286
References	289
Name Index	304
Subject Index	310

PREFACE

Every age in life is important and offers distinctive opportunities. This book explores some common questions about the human experience from conception to death. The usual reason for studying growth and development is to better understand people we love, work with, and serve on the job. Sometimes this means focusing on childhood and adolescence. We want to know what it's like to be growing up today, how this experience differs from our own upbringing, and what we can do to facilitate learning and wellness. We can become a source of guidance for youngsters by understanding current behavioral norms as well as principles of development that transcend time.

Our society is unprecedented in the need for self-improvement during adult life. We expect men and women to continue learning beyond the formal school years and behave in ways that allow a high level of self-esteem. These are worthwhile goals but cause us to sometimes wonder how our experience resembles and differs from other people. The notion that each person should feel unique promotes individuality but it may also lead to feelings of isolation. By knowing what others in our age group find satisfying and difficult, a peer standard can be considered in matters of self-judgment.

All of us are aging and considered to be an older person in the estimate of someone. Because a long life is probable we should prepare ourselves for the future by learning about the possibilities and limitations typically experienced in middle and later life. Besides determining the habits and skills we need for productive aging, such knowledge will also enable us to be more responsive to the needs of older relatives, friends, and colleagues. The theoretical background and practical guidelines offered here are intended to increase your knowledge, level of maturity, and constructive influence.

Chapter 1

UNDERSTANDING THE LIFESPAN

This book is about the continuously changing patterns of human development that take place from one age-stage to the next. The typical way to describe age-stages is to devote a chapter to each age in succession. Another possible approach is to select certain aspects of development—such as physical, emotional, or cognitive—and within each chapter show how the developmental patterns merge, flow, and evolve into progressively higher levels of functioning. This is the pattern chosen for use in our text.

In this chapter the (1) common methods for studying growth and behavior are discussed; (2) a concept of age-stages is presented with the parameters that are used throughout the book; (3) the major psychological viewpoints, often called "schools of psychology," are considered; and (4) some pervasive principles or developmental trends are cited. These principles are applicable regardless of the viewpoint adopted or the age-stage being considered.

Sources of Human Development Data

Forerunners of Current Knowledge

In ancient and medieval times data on human development came from the insights and perceptions of scholars and philosophers. Most of their theories have been discredited but some of the terminology survives; for example, phlegmatic, sanguine, melancholic, and choleric personalities. More recently, William James, William McDougall, Stanley Hall, and Jean Piaget initiated and extended contemporary psychology with their insightful perceptions. Sigmund Freud added to these fast growing ideas with insights and hypotheses which pervade current terminology. Specifically, we have from Freud such terms as introvert, extrovert, ego, various complexes (maternal, paternal), and theories of the nature of human nature.

Contemporary data stem from several kinds of systematically gathered and precise information. Brief descriptions of major methods follow.

Biographical Studies

The earliest systematic studies of childhood were baby biographies which were pioneered by Pestalozzi (1746–1827). The shortcoming of such studies is the almost irresistible temptation of the biographer not only to record, but actually to see, that which he wishes to see. Despite drawbacks, baby biographies have, at the very least, provided clues and informed guesses about the course of human development.

Questionnaire Studies

Questionnaires—a schedule of inquiries that is the same for all subjects—offer a way to limit the scope of observational studies. They may improve the margin of error but have the same limitations as empirical and biographical data, i.e., limited scope and observer error. The technique has the advantage of being able to focus on particular aspects of development (mental, linguistic, social). Data can be checked by comparing results from several subjects and surveyors. A schedule of things to be observed when looking through a one-way window is actually a type of questionnaire.

Cross-Sectional Studies

Cross-sectional methods form the bases for the bulk of studies on human development. In this approach researchers take measurements and make observations of many individuals of a given age or category (sex, grade in school, socioeconomic status). Averages are obtained from the results and trends are noted. For example, the norms of the most mental tests are derived from cross-sectional study. Establishing trends permits comparison of the individual with the hypothetical average for age, grade, or social status. However, such norms tend to obscure individual differences and have, on occasion, caused undue concern when individuals deviate slightly. Height-weight tables, found in typical textbooks on development a few decades ago, are seldom cited today. Researchers now recognize that differences are normal, that uniqueness is typical, and that cautious interpretation of measures of central tendencies is necessary. The "tyranny of the average" is a real hazard in understanding humans.

Cross-sectional data are steps forward in the science of human development. Such investigations are convenient for short-term studies. They yield attractive hypotheses and propositions. They include the phenomenon of variation, and used cautiously, they can help create awareness of abnormality.

Longitudinal Investigations

Longitudinal studies record observations and measurements of the same subjects (as far as possible in terms of death and geographic dispersal) over a period of years. Time-consuming and difficult for individuals to pursue, most longitudinal studies are institutional, conducted continuously by successive teams. An example is the Harvard Growth Studies which began in 1872 and continues to the present. Others pursuing longitudinal studies include the University of Iowa Child Welfare Research Station (1917) and the Yale Institute of Child Welfare of the University of California (1928). The Fels Institute of Antioch College (1929) has contributed studies of many aspects of child development and is particularly noted for discoveries about the conditions and factors of pregnancy, prenatal, and neonatal development.

Lewis M. Terman (1925) used the cross-sectional approach in developing the Stanford Tests of Intelligence. Subsequently he used the longitudinal approach in his Genetic Studies of Genius. These began by recording the background, health, character, and school achievements of gifted children in the San Francisco Bay area. Later the same persons were studied as adolescents and youth, and then a report was made on their mid-life status; careers, marriages, health, accomplishments. Some were quite ordinary and others achieved preeminently.

Experimental Methods

The basic ingredients of an almost endless variety of experiments are a hypothesis, and experimental group, a control group, a dependent and an independent variable. The experimental group is the one to which some set of conditions which accord with the hypothesis is applied (praise, rewards, drugs, physical setting, etc.) during pursuit of a stated goal. The control group (consisting in persons of the same age, socioeconomic status, health, school grade) is not subjected to the special conditions of the experiment but is given the same tests regarding achievement of the goal. The special conditions of the experiment are called the independent variable (praise, rewards, etc.) The outcome of the experiment (improved performance, level of scores) is called the dependent variable. The comparative results on the experimental and control groups are often subjected to correlational or other statistical treatment.

Clinical Studies

A recent addition to data gathering approaches is called the clinical method. This recognizes the need for getting beneath, and beyond the averages, trends, and probabilities (gathered by the procedures described above) to the specifics of a given person or case. Some persons claim, with considerable justification, that there is no typical or general adolescent, or child, or three-year-old, middle-aged person, or average old person. It is necessary to know why *a given individual* acts as he does. The subject of a clinical study in medical practice is given a variety of tests and questionnaires. Doctors with various specialties contribute their knowledge to reach a consensus. A child in a reading clinic (or behavior clinic) is given tests, is observed, interviewed, and has his life history (health, school record, family) recorded. These data are combined by a number of consultants who try to explain, diagnose, and make recommendations in terms of all the clinical information for one specific developing person. Clinical studies are a team approach to understanding one person.

Using Data on Development

Because any one method of gathering data is limited in some way and because some methods are particularly useful in gathering data on specific areas of development, our cumulated fund of data is the result of many combined approaches. In addition, methods that were once thought to be adequate, e.g., mental testing, are now being questioned and widened sources of data are being pursued. Again, for example, mental testing is now being supplemented with studies of lifestyles and day-to-day performance.

It is probably true that we are only on the threshold of understanding human development and behavior. We are, nevertheless at a point where the application of what is currently known can, if applied, vastly improve the lives of humans (e.g., innoculations, enforcing drunk-driving laws). Readers might pause and enumerate some of the desirable applications of that which is known but not yet in widespread use.

VIEWPOINTS ON DEVELOPMENT

What one sees both scientifically and in the everyday world depends upon what one is seeking. The grouse hunter may be startled by the snort of a deer that he has approached closely. If he were hunting deer, the roar of grouse wings would be the heart thumper. The seven-year-old boy is likely to see in a poster the football player; but the adolescent's eyes widen at the picture of designer jeans. Psychologists, despite their pleas for objectivity, readily find support for their favorite theory. In this section we cite some of the major perspectives on human development.

Hereditarian Perspective

In 1940 the National Society for the Study of Education published a two-volume yearbook on *Intelligence: Its Nature and Nurture*. In essence authors of the yearbook declared that both heredity and environment contributed in unknown amounts to intelligence. Our job was to use and develop these potentials with which we were born. This advice was widely accepted and for a number of years we rested comfortably with the idea that intelligence could be nurtured by wise management of environment. Then Arthur Jensen (1969) wrote the longest article ever published in the *Harvard Educational Review*. The results of his search startled psychologists and educators with alleged proof that we might as well forget environment (including preschool and kindergarten) as far as mental development was concerned. So much of intelligence is inherited, said Jensen, that environmental intervention was futile. In more recent writings, including his book *Bias in Mental Testing*, Jensen (1980; 1984) reiterates his contention. In this text our bias is that environment is important and early intervention can be extremely helpful in the development of human potential.

Such things as eye color, bone structure, and hair texture seem to be inherited, but these are not behaviors. Humans, at most, inherit only the potential for developing certain kinds of behavior. The pattern of human fingerprints is probably genetically determined. We can conclude at present that hair color is inherited (red-headed Africans who suffer protein deficiency excepted). A number of defects may be listed in which environment seems to play no part. These include absence of hands or feet, defect or absence of teeth, fragile bones, long lower jaw, excessive body hair, hollow chest, poor muscular coordination, soft dentine, progressive deafness, and colorless skin spots. The foregoing are dominant hereditary traits. In addition, some thirty recessive traits are noted as being little affected by environment, including certain types of mental defect, albinism, sickle-cell anemia, baldness, color blindness, hemophilia, eye tumors, and other eye defects. Heredity has its greatest influence on physical traits. Thus, genetic factors play the major role in facial appearance (size and shape of jaw, lips, forehead, and nose), in body build (height, bone structure, and body proportion), and in many physical abnormalities.

It appears there is much that heredity does influence even after denying its power to shape behavior. There is a constantly growing body of information about the nature of heredity and its input on features and behavior potential. Knowledge of genes, chromosomes, and the "building blocks" of heredity, DNA, is making possible the science of human engineering (Felsenfeld, 1985). So far our knowledge of such engineering is limited to the prevention of defects. However, from the outset some of the leaders in this field have expressed considerable hesitancy about its application. Glass (1969, p.

506) said there is the distinct prospect that genetic theory ". . . places in human hands the possibility of henceforth modifying all life, including the nature of our own species." Howard and Rifkin (1978) acknowledged that genetic engineering might eliminate cystic fibrosis, Down's syndrome, and sickle-cell anemia but that the required decisions must for ethical reasons be approached reluctantly and cautiously. A similar warning is offered by current leaders in genetics and futurism (Harper, 1985; Naisbitt, 1982; 1985).

The hereditarian viewpoint gives hope for the diminution of the incidence of defect but none thus far for the enhancement of potential. For practical purposes and certainly for those already born, the hope for optimum development resides in the physical and human environment.

Behavioristic Perspective

Many of the early psychologists—McDougall, James, Freud (actually a physician rather than a psychologist), and Jung attributed much of human behaviors to instinct—an inherited mode of response. Decidedly the majority of developmentalists attribute little of human behavior to instinct. Rather they resort to words such as predispositions, urges, needs, and appetites, and learning (Tomlinson-Keasey, 1985).

Several psychological orientations arose in response to and rejection of the instinct, inborn hypothesis. The first to gain rather widespread popularity was the behavioristic viewpoint espoused by John B. Watson (1919). He postulated that there are few instincts (fear of falling and loud noises, rage, and love); and that humans learned through experience to be what they were and might become. With planned intervention one could make of newborn persons that which the manager of the environment sought. He said, "There are inheritable differences in structure but we no longer believe in inherited capacities. Give me a dozen healthy infants and my own world to bring them up in and I will guarantee to train them to become any type of specialist I might select—doctor, artist, merchant or chief, beggarman or thief" (Watson, 1930, p. 104). He did succeed in teaching babies not to fear dogs, rats, snakes, or darkness. However, he later modified the claim about making children into chiefs or thieves. He shifted his terminology to the need for starting with normal children.

Behaviorism in its fall from popularity was criticized on several issues: (1) by Gestaltists for oversimplifying complex human behavior; (2) by humanistic psychologists for regarding persons as some sort of machines (it was regarded as a mechanistic view); and (3) by proactive psychologists for not recognizing the input of the behaving subject.

Behaviorism still has its staunch adherents—and with good reason. It works. Skinner (1954; 1971; 1983) acknowledges the complexity of behavior—including its social context. But he has great faith in the modifiability of behavior. He is perhaps best known for his endorsement of behavior modification and its psychological component, operant conditioning.

Behavior modification refers to the phenomenon of teaching/learning those actions and conducts that will be effective in living productively with one's fellow beings. The subject—child or adult—is conditioned to act constructively by using planned stimuli and systematic reinforcements. The reasons why children misbehave is that parents and teachers do not realize what they are reinforcing. For example, the well-behaved child, playing quietly, may be smiled on contentedly by parents but he is not given overt attention. The child throwing a temper tantrum gets all kinds of attention—some of it may be physical but it is better than being ignored.

Operant conditioning refers to the giving of rewards (a kernel of corn to a chicken, a piece of candy to a child, a word of approval or a smile to an adult) for conduct in progress; i.e., action in operation. To illustrate, the boy who hangs up his coat is praised for his action; at least it is obviously noticed, "I saw you hang up your coat. There's some orange juice in the fridge if you want it." The reinforcement comes from that action in progress which is sought, or approved by, the experimenter—or conditioner. Operant conditioning works: in reducing temper tantrums, in getting students to apply themselves to arithmetic problems, in helping people to stop smoking, in getting husbands to help with the dishes.

Gestalt-Field Orientation

The gist of Gestalt psychology is contained in the word itself. Gestalt is a German noun (capitalized because all German nouns are capitalized) which means approximately shape, form or configuration. The emphasis in Gestalt psychology is not on such a simple thing as S → R (stimulus → response). The idea is that behavior occurs in a configuration, a context, a total environment. For example, he who was yesterday the life of the party behaves quite differently in church, or when he walks down the aisle at his daughter's (or his own) wedding. Children at home often behave quite differently than in school. The Gestaltists stress the contextual nature, or totality, of behavior by emphasizing figure-ground. The figure is sharp, distinct, small, and rather clear; the ground is amorphous, vague, massive, and without boundaries. Teaching is to a marked extent a matter of making sharp and clear some of the things that are enmeshed in the child's perceptual ground. Together figure and ground make up the field; hence, the term Gestalt-field psychology.

To the Gestaltist learning is more than a matter of exercising stimulus response bonds. One might also learn through perceptual reorganization. This reorganization is called insight—an "aha" experience, or "Oh, I see now." Two of the founders of Gestalt psychology Kurt Koffka (1935) and Wolfgang Kohler (1925) developed the concepts of insight, behavioral field, and motivation by experimenting with apes. In one experiment a caged ape tried to grasp a banana just beyond the bars. He sat in a corner and saw a stick inside the bars which was in line with the banana. He took the stick, "aha," and fished the banana within reach. Insight! There was no repetition of futile reaching, no trial and error—just quiet contemplation of banana, a stick that emerged from the amorphous background as a tool, and insight. The behaving organism is part of the field and the insight occurs within the organism.

Kurt Lewin (1936) used Gestalt principles in translating the theory into experiments and explanations of human social behavior. His experiments in social climates—authoritarian, laissez-faire, and democratic—have become classics in psychology. He was one of the first researchers to call attention to cross-age influences on behavior and personality. Under authoritarian adults, preadolescent boys became rebellious, aggressive, destructive, and projected their frustrations on scapegoats. Under democratic leaders—those who listened, discussed, and suggested—boys were cooperative, purposeful, wanted to remain in the group and considered "oncoming generations," (the boys who would next occupy the room). Teachers, parents, juvenile authorities (and yes, wardens in penitentiaries) are beginning to take note of these insights regarding social climates (McGrath & Kravitz, 1982).

The process of introspection had been disdained and discarded by behaviorists because it was subjective. But Lewin (1948) used introspection to offer some valuable clues for understanding racial prejudice in his report of what it is like to be a Jew. As our society faces ethnic confrontation and as parents and teachers when we scold, shout, blame, or threaten children, it might be well to reflect on Lewin's conclusions. Specifically, (1) members of minority groups and victims of blame come to accept the view of the majority; (2) much of the potential of minority persons is lost via chauvinism; (3) we will editorialize: Children who are told that they are dumb, disobedient, and unworthy also internalize such evaluations.

Snygg and Combs (1949) used earlier psychologies, including Gestalt, to emphasize such things as the three items just mentioned. Their approach was called phenomenology; i.e., human experience consists in how individuals see themselves and the parts of the physical and social world that have significance for them. Let us look at item #3 above: Children do not see themselves as being poor, helpless, or underprivileged—if their parents love and care for them—at least until children attend school. They do not see themselves as being stupid, burdensome, or alone if they are not ignored, blamed, or battered. Behavior, say the phenomenologists, is an individual matter because it is the phenomenal self that sees the external world.

Psychoanalytic Orientations

Psychoanalysis was formulated by a medical doctor, Sigmund Freud, who, despite his training in a biological science, came to see that many human ills were psychological. He perceived behavior as being determined. This determinism was powered by biological factors, social and economic inputs, and numerous other factors as it shaped adult personality. Roberts (1975, p. 8) suggests, "As the psychology which is most dominant in our culture, outside of academia, Freudian psychology provides the major ways we think about ourselves and our culture."

The influence of Freud (1925; 1935; 1962) has been significant for a number of years but it should not be overstated. We think there may be many nonprofessionals who endorse behavioristic interpretations. We can agree that psychoanalysis was, and is, dynamic, changing, and influential. Certainly some of the terms Freud used have earned a permanent place in psychological writings and daily discussion.

By far the major part of personality (in psychoanalytic theory) is the id, which is unconscious; it is the source of the driving force of life as it seeks pleasure and expression. The instinctual avenue to pleasure is the libido or sexual energy. However, the ego, which is in contact with reality, blocks the way to free expression of libidinal urges. Just before libidinal forces surface they encounter the preconscious, which acts as a censor to the primitive drives and attempts to mold the drives in ways that make them compatible with the ego and its sense of reality. The ego is aware of the superego (the conscience) and the biological urges; so attempts are made to find ways to express libidinal urges within the approaches approved by the conscience. A fundamental factor in psychoanalytic theory is polarity—one force or drive against another. Polarity is found, for example, in life versus death wishes, love versus hate, autonomy versus independence. If the ego does not find a solution for such conflicts, it seeks bypasses, detours, and evasions through the use of defense mechanisms. The general public may not refer to the id, the libido, or the superego but is aware of conflicts and defense mechanisms. They can identify rationalizations, repressions, projection, identification, psychosomatic illness, and hypochondria—each term a legacy of Freud.

If it seems that ego, id, and superego are referred to as live beings, the reader may get some reassurance from Freud. He animated these drives with names from Greek mythology. He referred to the Electra complex, which is the adolescent girls' desire to have sexual relations with her father. He spoke authoritatively about the Oedipus complex which is the urge of the adolescent boy to be husband to his mother. "How absurd," some of you say. Freud has still "got ya." Your ego has used the defense mechanism of repression—you are denying your own instinctive urges to please your conscience. (We don't believe it either.)

Much of the vocabulary of psychoanalytic theory has been accepted into the mainstream of psychology; e.g., ego, defense mechanisms, polarity, and personality conflict. In this book considerable attention is devoted to the ideas of a psychoanalytic adherent, Erik Erikson (1963; 1980). He contends that life consists in a continuous series of crises, or critical periods as persons from birth onward attempt to resolve the conflicts which arise from the polarities encountered in normal daily living.

Humanistic Psychology

Humanistic psychology has its roots in the philosophy of humanism—which reaches back at least to the time of early Greek history. It calls attention to the worth and centrality of individuals. Abraham Maslow (1970), who introduced humanistic psychology, is best known for his theory of motivation, his emphasis on self-actualization, and his focus on healthy persons. He repudiated those psychologies which sought to gain insight into normal behavior by studying sick and maladjusted persons—as was the mode up to mid-twentieth century. He contributed much to all of us by his investigations of the values, behaviors, and motivations of those who were unusually healthy; i.e., self-actualized persons.

Maslow says (1962, p.21), "Observation of children shows more and more clearly that healthy children enjoy growing and moving forward, gaining new skills, capacities and powers." This may not be true for insecure, frightened, or sick children; but it is true for those who are healthy. Growth motivation is a pivotal factor in human development at all ages. Rogers (1983) called it directional growth and likened it to the tip of a tree—ever thrusting upward. If the treetop is broken off, the next lower branch will turn upward and become the leading part. Menninger, Mayman, and Pruyser (1963) referred to the growth motive as the heterostatic urge. They contrasted it to homeostasis, which is the desire of the organism to come to a state of equilibrium or rest. Healthy and rested individuals seek activity and challenge. Menninger wrote from a psychoanalytic persuasion.

At one time it was believed that the essence of cognitive development was problem solving. Now it is postulated that a few optimally developed persons are not satisfied with solving dilemmas—they actively search for problems and difficulties. This high stage of cognitive development is called problem finding (Arlin, 1977). There are some people, who, far from being satisfied with a state of physical equilibrium, are active sensation seekers. They look for mountains to climb, and, like Jacques Cousteau, new worlds to conquer. Such too are the well-paid corporation executives who take dollar-a-year jobs in government. They seek the challenge of public service and social problem solving.

Charlotte Buhler, a prominent advocate of humanistic psychology, perceived life as being lived through progressive, sequential stages. Her (Buhler, 1968) emphasis on goals and their constant, or at least periodic, revision reflects a basic tenet of humanistic psychology: progressive individual change. Buhler's views are of interest because she shows how choice, responsibility, identity, self-esteem, and interpersonal relations change and develop through various age-stages of life.

Proactive psychology is usually considered an emphasis within humanistic psychology. According to Bonner (1965) humans are inherently both reactive and proactive. That is, one does behave reactively to external stimuli; but, also, one has input into the psychological milieu (is proactive by virtue of his perception of the world). Thus, one's lifestyle can be one in which there is choice and responsibility; or, the prevailing lifestyle may be one of reactivity. Both psychoanalysis and behaviorism ". . . are fundamentally reactive psychologies—psychologies which define behaviors in terms of stimulus and response framework" (Bonner, 1965, p. 48). There is, says Bonner, a creative, spiritual, moral nature of man that if not recognized, makes the study of human nature and human development futile.

The underlying postulation of humanistic psychology which began in the early 1960s was that if the developmental circumstance of growing persons can be essentially positive and salutary then maturing persons can realize their inherent potential, live congenially with themselves and others, and become creative, appreciate beauty, and simultaneously exercise their logic and intelligence. They may become fully human.

Social Learning Theory

Proactive psychology seeks to emphasize that individual input, individual interpretation, and individual goal pursuit must not be minimized. Social learning theory seeks to keep today's students and theorists from understating the fact that stimuli occur in a context, especially a social context. Cognitive development stems from the involvement of many persons and stimuli acting simultaneously on the developing person. Bandura (1977) warned against placing too much emphasis on laboratories with their mazes, puzzles, and conditioning experiments. Direct observation of subjects in unstructured situations is time consuming and difficult to analyze and interpret but is fundamental to accurate understanding. Focussed study of persons in unstructured environments can reveal clues that accumulate to provide data on how children imitate, experiment, and use models as they develop behaviorially. Bandura (1977) contends that there are specific principles of learning involved in such acts as children's imitating and observing. There are identifiable processes involved in children's verbal milieus—which can later be tested in experimental situations.

Social learning theory does not ignore nor derogate behavioristic or humanistic psychology. It adds to them. It adds to itself (i.e., social learning theory is not a static concept). At present salient features of social learning theory are:

> Development is a two-way interaction between self and environment.
>
> Persons may learn through observation without reinforcement.
>
> A subject's expectations and perceptions influence what he chooses to do.
>
> Persons are enactive processors of incoming data rather than machines or robots incapable of self-evaluation and choice (Helms & Turner, 1986).

Information Processing

Theories arise out of the need to consider increasingly large amounts of data as more detailed information about developmental nuances are recognized and still keep explanations simple and systematic. For example, it was once thought that memory consisted of mental traces of vivid or repeated stimuli. As knowledge grows and nuances are discerned, it is realized that memory is more than addition or repetition and vividness. It involves a reacting and proacting organism. Thus, a recent supplement to developmental data has come to be known as information processing.

Gagne (1977) has designed a schematic representation of information processing in cognitive development. External stimuli are received by sensory receptors. Those stimuli that are accepted become transformed in a fraction of a second to neural activity (visual, aural, or verbal) and are retained as short-term memories. Those neural activities that are significant to the individual are sorted and encoded and become a part of long-term memory. Some of these, in appropriate situations, are retrieved (some readily and some not so readily). All these interactive and reversible processes culminate in the generation of a response—physical, linguistic, mental. These responses are confirmed or allowed to lapse depending on the feedback from self or others. Behaviors getting positive feedback tend to be consolidated and become more or less stable learnings. Gagne has used the information processing model to test, confirm, and restructure step-by-step classroom learning-teaching experiences.

DEVELOPMENTAL AGE STAGES

Part of the rationale for the organization of this book (discussing all ages within every chapter) is that although development is continuous, the needs of each age-stage are somewhat different. These distinctions are seen more clearly with a lifespan approach than is the case where separate chapters are concerned with each age stage. An introductory word should also be said about the definition of age-stages. There is variation among authorities in terminology so we have defined the stages somewhat arbitrarily but in agreement with some consensus. These are:

> Ancestry—prenatal 9 months (hereditary and congenital factors)
> Infancy—birth to age 2
> Early childhood—ages 2 to 6
> Middle childhood—ages 6 to 9
> Later childhood—ages 9 to 12
> Adolescence—ages 12 to 20
> Early adulthood—ages 20 to 40
> Middle adulthood—ages 40 to 60
> Later adulthood—age 60 to death

Ancestry

The word ancestry is used here instead of heredity because it implies *people* more pointedly than does *genetics*. New discoveries about the dynamics of heredity are being made, especially as genetics becomes a mathematical and computer science, rather than relying on experiments with peas, mice, and fruitflies. As more is discovered about DNA (deoxyribonucleic acid) and RNA (ribonucleic acid) the more prominent human intervention in working of ancestry become. The intertwining (actually as well as figuratively) makes it possible for genetic engineers to offset or avoid defect. Those who frown on humans tampering with God's right to control heredity must contend with the argument that we have, throughout history, interfered with hereditary potential by devising protective developmental environments (Darnell, 1985; Felsenfeld, 1985).

There are several salient prospects in these discoveries:

1. Carriers of grossly defective genes (previously known only by result) may be identified and sterilized or institutionalized. However, there are so many carriers of defective genes that this would be impractical.
2. Genetic surgery (for example, using laser rays) may be possible in the future. This would mean reaching into the just-fertilized egg and altering the defective gene.
3. Prenatal adoption is possible; implantation of an embryo in the uterus of a surrogate mother has already been achieved.
4. It is possible to test and type prospective marriage partners for recessive, defective genes and then to decide whether or not to marry, to remain childless, or to use adoption, either prenatal or postnatal.
5. The study of genetics and molecular biology may have implications for learning research as the chemical action of the brain is analyzed (Harper, 1985).

Infancy—Birth to Age 2

The term infant comes from Latin and means incapable of speech. For many years the scientific study of infancy was overlooked precisely because babies could not orally report what they saw and understood. The language restriction also led adults to believe that not much learning took place before speech is acquired during the second year. Another factor contributing to the underestimate of early learning was a common lack of memory by adults of their own experience as infants. If this age-stage was important it would surely be remembered by at least some of us. Instead, because we all seem to have forgotten infancy, it must not be a particularly critical period for intellectual growth.

Fortunately, in the late 1950s some innovative research methods led to more accurate impressions about infancy. While babies were shown two different objects their eye movements were observed. Measurements were taken regarding what the infants look at and for how long. It was found that they prefer complex designs to simple ones. Before long there was general agreement that by combining the use of videotape and precise observation a broad range of nonverbal behavior could be assessed. As ever younger infant groups have become the focus for investigation, the level of respect for their intellect has continued to rise (Brazelton, 1983; Friedrich, 1985; White, 1985).

Although babies are more capable of thinking than was once supposed, they appear to be more emotionally vulnerable too. Pediatricians Klaus and Kennell (1983) direct a hospital intensive care unit. Despite their efforts to save premature children, a disproportionate number of those who were sent home in good health showed up again at the emergency ward as victims of abuse. It seemed that the traditional practice of prolonged separation of mothers and infants in cases of premature birth might be a contributing factor. The emotional attachment between infant and mother, referred to as bonding, can be interfered with in cases of separation. To test their assumptions, Klaus and Kennell permitted mothers to enter the nursery for premature infants even though other physicians opposed the idea erroneously supposing that babies would be contaminated. After several visits mothers moved from reluctance to enthusiasm and desired more closeness.

The need for mother and infant to be together initially has also been demonstrated in normal delivery situations. One group of mothers and babies received regular care while another group were placed in a warm room where mother and infant lay together unclothed for an hour following birth. The early contact pairs were united five hours a day during each of the hospitalized days. Interim results of a five-year follow-up study show that the early contact mothers interact more with their children, show them more attention, and more frequently provide affection. They issue fewer commands, ask more questions, listen more attentively, offer more feedback, and model better speech. Although both parent groups were similar in socioeconomic background, the early contact moms presented a model of better learning. Their children have a greater opportunity to develop trust and confidence, essential ingredients for intimate relationships and mental health.

Early Childhood—Ages 2 to 6

One reason for the widespread attention given to early child education comes from studies indicating that as much intellectual growth occurs before children enter kindergarten as during the remainder years they will spend in school. If general intelligence is a developmental characteristic related to the time it takes to learn skills and concepts, it would seem reasonable that lack of learning in the early years might be difficult or even impossible to remedy fully at a later time. On the other hand, excellent learning during early childhood is unlikely to be lost. From this view of intellectual growth has come the persuasion that we must find ways to improve the environment of preschoolers so that as adults they will have greater intellectual ability than they would otherwise. It was precisely this goal—to provide a better preschool experience for children from deprived homes—that led to the establishment of the nationwide Head Start program twenty years ago. Personality as well as intelligence is being formed during this age stage. For both kinds of development, the question asked is usually the same: How important is the influence of environment? Actually, environment can effect people of any age but the younger the person, the greater the impact. Therefore, the younger the children, the more influential those who teach them can be (Bronfenbrenner, 1984).

Who is responsible for teaching preschoolers? During the 1980s a rising percentage of mothers entered the labor market. They want access to high quality, economically attainable preschool education for their eleven million children. Because many of these families are middle income, society has been slow to respond to their pleas for help. Much debate also focuses on preschool personnel, particularly their usual lack of training, ratio to children, high rate of turnover and low wages. There is also controversy about what to expect of children. Here the guiding question has recently changed from "What are boys and girls capable of learning?" to "What kinds of learning are best during early childhood? This shift in concern is overdue, according to many kindergarten teachers whose disappointing experience with forcing academic goals led them to seek assignments at a higher grade level. Meanwhile, parents want a place to send their child where the caretakers will cause them to experience self-esteem, learn to get along with age-mates, enjoy learning, and develop readiness for elementary school (Carnegie, 1985).

By honoring the ways young children prefer to learn and accepting the limits of their age stage, parents and teachers can make instruction more worthwhile. Because play is the dominant activity of preschoolers, their favorite way of learning, some researchers have attempted to determine which child-rearing goals can be achieved through this medium. A valuable by-product of these investigations is new knowledge about the process of play, especially those factors that inhibit and support satisfying interaction. Some remarkable changes are taking place in attitudes. Whereas parents once felt an obligation only to read to preschoolers, they now consider participation in fantasy play to be another aspect of their responsibility. Parents' growing realization that their children need to play with them as well as with peers and that the benefits of parent-child play are different from those offered by age-mates has created a new demand for guidance. Mothers and fathers want to know how they can use toys and televiewing to help children develop language competence, values, creativity, memory, and problem solving skills. In short, they have become interested in teaching through play. Although this transition from indifference to enthusiasm toward parent-child play is a favorable sign, the motivation for involvement should go beyond instruction to include mutual enjoyment and gains in the family relationship (Baldwin, Colangelo, & Pittman, 1984).

Middle Childhood—Ages 6 to 9

There are several considerations that give this age-stage distinctiveness. Probably the most notable is the accelerated move toward autonomy. Primary grade children have an urge to get out from under the protection and direction of their parents. There is, however, ambivalence about this motive. They want autonomy but they don't want it; consequently, many youngsters turn to teachers with respect and a readiness to obey and believe them.

In middle childhood many are using symbolic thinking and imaginations are active. Absurdities and jokes are typically enjoyed, causing bursts of laughter out of all proportion (in adult eyes) to the size of the joke. However, it is necessary to note that such generalizations are generalization—not rules or schedules of development. A growth principle might well be derived from normal variations: "The older persons become the greater the differences between them becomes." To be sure, we have an exception. While some children tend to seek autonomy, others are slavishly molded by the need for peer conformity.

The psychoanalytic view postulated by Freud (1962) is that middle childhood is a period of latency; i.e., the id is relatively quiescent and problems of sex interest and sex identity are minimal. Erikson (1980) views this as the critical period of industry versus inferiority. Depending on how well they perform expected tasks children will see themselves predominantly as admirable, industrious, and productive workers or as inferior and inadequate. Because children's views of themselves reflect the views of their caretakers and associates, adults might appropriately be advised to see good in children, use praise liberally (but not inappropriately), and to put on the rose-colored glasses that see beauty in vacillating behavior.

Behaviorists perceive middle childhood as being a time of responsiveness to those learnings which occur at home and school. For example, there is no latency in children of lower socioeconomic status who see sex less inhibited than in the upper socioeconomic status—in which Freud made his observations (Goldman & Goldman, 1982). Social learning theorists perceive reinforcement as being essential to the development of self-concept, pride in work, and socialization (Bandura, 1977).

Cognitive theory emphasizes the tendency in middle childhood to move away from self-centeredness in both thought and behavior (Gagne, 1985). Play and games with rules and protocol are important in the social and moral development of children. The loud group arguments that occur in recess sports are for American children normal and desirable aspects of development. Adults who are disturbed by the bickering need to know the impetus to socialization that underlies the contentions.

Later Childhood—Ages 9 to 12

The final stage of childhood is sometimes called preadolescence. During this timespan there is an acceleration of physical growth. The boy who has gained only two or three inches in height during his tenth year, may in his eleventh year gain 4, 5, or 6 inches. Girls too experience this acceleration—on the average, a year or two earlier than do boys.

A particularly noteworthy item is not so specifically related to later childhood. It involves the way age-stages are divided and the reasons for such division. Halfway through a delineation of age-stages we are only up to age 12. This concentration on the early years is because changes take place so rapidly. Many of the themes of development, or lifestyles, are being established in these early years and the effectiveness of intervention is greatest during the period of most rapid growth. Later childhood is the last age-stage in which physical change and cognitive development is so rapid. Adults should be reminded that efforts during childhood to enhance ego concepts, sense of responsibility, and moral standards will probably pay greater dividends than will efforts at a subsequent time.

Later childhood has been called, by those who have made it a concern among life stages, a virtual no-man's land in terms of research investigations. This neglect has been somewhat corrected in recent years (Juhasz, 1983). These studies show a number of characterizations of this period:

1. There is continued slowing of growth from middle childhood until toward the end of preadolescence, when the "prepubertal spurt" occurs.
2. There is a perceptible increase in strength, manual dexterity, and resistance to disease.
3. Interests and activities shift from child dependence to stabs at independence.
4. Normally there is an increase in parent-child conflict because parents have difficulty in terms of their daily contact perceiving the cumulative impact of growth processes.
5. The thrust for autonomy and independence from parents make the peer group highly important.
6. Physical and cognitive skills provide the basis for group membership, identification, and activity.
7. The typically more rapid rate of development of girls over boys is especially apparent in the late childhood years and promotes marked differences in the social orientation of the sexes. These average differences between boys and girls are not sufficient to provide differentiation of school curricula (Mussen, 1984).

Adolescence—Ages 12 to 20

Adolescence has been called a Franco-American invention and for good reason. The term has its roots in culture more than in age-stage. G. Stanley Hall (1904) correctly perceived the cultural ramifications of this period as is shown by the full title of his book, *Adolescence: Its Psychology and Its Relations to Physiology, Anthropology, Sociology, Sex, Crime, Religion, and Education*. Hall defined adolescence as starting at puberty and continuing until adulthood—both of which began at various ages. He saw sexual development as a part of the turmoil which supposedly characterizes this period. Freud (1925) added repression, aggression and defense mechanisms to the list of causes explaining adolescent behavior. Thus arose the rather strongly entrenched misconception that adolescence was a time of stress, confusion, alienation, and rebellion.

Fortunately, the work of Margaret Mead (1928) and Ruth Benedict (1935) initiated the opposing idea that culture rather than physiological and cognitive changes were what caused adolescence. Puberty is universal but adolescence is not. The notion that it is a Franco-American invention recognizes that industrial-technological society is what causes adolescence and much of its alleged stress. Specifically, Mead and Benedict found that in some cultures, where pubertal ceremonies moved children literally overnight into adult status, responsibility, and privileges, there was no adolescence. And, while we are challenging a conventional notion, let us also state that the majority of adolescents go through this time of life quite serenely, confidently, and competently—certainly with no more stress than is normally encountered in other age-stages.

Adolescents correctly feel that barriers have been erected which delay their entry to adulthood. It is admitted that some of these barriers were placed with the welfare of the adolescent in mind:

Extended education. Schooling leads to great opportunities but its increasing length delays the time for achieving financial independence, personal autonomy, and social identity.

Objections to marriage. Resistance to early marriage exists despite the fact that males reach the height of their sexual urge and power in their late teens and females are fully responsive at this age. Statistics, however, underscore the wisdom of waiting to marry.

Demographic influences. The process of defining adolescence is a persistent problem. In the nineteenth century our nation was young. Even today one half of the world's population is under 15 years of age. But as longevity has increased in the United States the median age rose. The result of this is "Lacking a meaningful ideology about what adolescence should mean in our society, we may fail to provide a sense of purpose and structure for growth at the very time in their lives when young people are seeking definition of their personal and social status" (Lipsitz, 1980, p. 25).

Rapid change. Some observers have noted that the current greatest threat to human life and development is rapid change (Toffler, 1970; 1980; Naisbitt, 1982; 1985). More dangerous than nuclear threat is the difficulty of achieving adjustive techniques and a sense of stability in the face of rapid change. This is a threat in all age-stages, once the age for logical thinking has been reached. Adolescents are particularly hard hit by rapid change because it is the time for them to be thinking about work, marriage, higher education, social responsibility, and personal identity.

Early Adulthood—Ages 20 to 40

The same difficulty of defining early adulthood is encountered as in the case of adolescence. Individuals of the same chronological age may be at quite different levels in terms of behavior, competencies, and task achievements. Society has removed some of this iffiness by stating certain legal definitions: age for voting, drinking, marriage, and being legally liable for debts. Uncertainty still exists because these definitions vary from state to state.

For convenience of academic discussion, early adulthood begins when a person assumes financial responsibility, at least for oneself; it continues as one becomes responsible for spouse and children. Birren (1970) epitomizes the approximate nature of age-stages by referring to functional age; i.e., individuals assume adult roles at various rates and ages.

Levinson (1978) concluded as the result of a longitudinal study of forty subjects that four main tasks are faced in this age-stage:

1. Defining the dream of what one wishes to accomplish in life.
2. Finding mentors who train, guide, and encourage one in moving toward the dream career (Torrance, 1986a).
3. Developing and exercising the skills necessary for career satisfaction. If there is ambivalence or uncer-

tainty about defining the career dream, one may become inefficient and develop anxieties.
4. Forming a marriage and family. Persons need the pleasures and responsibilities of parenthood for moving fully into the adult world.

Middle Adulthood—Ages 40 to 60

To the young person middle age must appear to be misnamed, inappropriate. Those years from 40 to 60 are well toward the end—not the middle—of the lifespan. Perhaps with increasing longevity the designation will become more appropriate. The word *middle* is quite accurate in terms of generational placement. Persons in this group are the mediators, the communicators, the middlemen, of cultural transmission between the young and all those who have gone before. We hypothesize that no other human trait or other cultural characteristic can transcend that sought in the middle years; i.e., the trait of generativity. Throughout this book we will explain and extend the meaning of generativity in various contexts. Here, generativity refers to the ability of functionally mature adults to be involved with the welfare of younger generations.

Stieglitz (1952) was among the first to highlight the significance and satisfaction of middle age. He perceived, as we do, the inseparability of the generations and the persistence of personality patterns—both in individuals and families. He regarded it as being absurd that adolescence should so consistently be seen as a preparation for adulthood but that middle age should so consistently be ignored as a preparation for old age. Stieglitz advised repudiation of the fallacy, "You're only as old as you think you are." He advised that physical changes, energy levels, diminished healing and recovery rates, and moderated ambitions be acknowledged and internalized. "Act your age."

Neugarten (1980) views middle age as a time of opportunity rather than crisis or decline. She regards middle age as a new prime of life because parental roles are less stressful; homemaking is less burdensome; spouses can have more time together; work roles are relaxed—performance rather than prestige is sought; citizenship shifts from self to other-concern; people become more important as children make fewer demands; leisure time is increased.

Several studies of middle-aged couples found satisfaction rather than crises in midlife. Lauer and Lauer (1985) report that marriage is as satisfactory in the middle as it was in the early stages. Many of the early parenthood problems had vanished. In terms of high to lower satisfactions the ratings were:

> Being a spouse
> Provider
> Homemaker
> Religious activities
> Community service
> Social activities
> Affiliations with professional organizations
> Educational roles

There are aspects of intellectual decline in some mental functions but continued increase in others. However, the evidence is clear in one respect: to the extent that cognitive powers are exercised the rate and amount of decline is lessened. Capacities—physical, mental, emotional, social—must be exercised if they are not to atrophy. The longer people live the more time they have to think; and in the process of thinking they may grow.

Later Adulthood—Ages 60 to Death

Despite the fact that the longer persons live the more different they become, we can quite arbitrarily define later adulthood. In a work-oriented culture, later adulthood is that age-stage which begins at about retirement and ends with death. Some persons have preserved their physical status and strength so they are in better shape than others who are 15 to 20 years younger. Some are as mentally alert and active as those who are still employed. Research on aging appears to be much more practical when it focuses on what old persons might be than when attempts are made to generalize on the wide variations of what is. This wide variability is reflected in the diversity of several theories of aging.

Physiological deficit theory. This refers to the wearing out of the organism. Metabolic wastes accumulate and result in excretory difficulties, brittle bones, hardening of the arteries and reduction of efficiency in metabolizing blood sugar and immune reactions (Butler & Gleason, 1985).

Genetic Theory. This view postulates that aging is a species-specific factor of inheritance. Families have markedly similar histories of being short or long lived. Stories about societies that have many persons 140 years old do not stand up to close scrutiny. However, in terms of our present genetic potential, nutrition, sanitation, and conquest of disease, 90–100 years is widely attainable (Gale, 1985).

Disengagement theory. This is the idea that later adulthood involves a progressive and often quite satisfactory withdrawal from such areas as career, social, and family roles. One becomes less concerned about health and finances. One seems to have come to terms with many of the exigencies of life. There are lively objections to disengagement theory because there are so many who simply do not fit the concept. It seems that engagement is the source of satisfaction just as frequently as disengagement. Successful aging depends not so much on what is done as who is doing it, i.e., the personality of the aging person (Teri & Lewinson, 1985).

Activity theory. This theory proposes that the needs of senior adults do not differ markedly from people in middle age. Those in later adulthood must keep up the good work or find some work-like alternative. Activity theory is illustrated in a model community, Lindisforne. There young and old are given vital complementary roles. It is claimed that society cannot afford to waste the "fantastic assets of old people" (Thorp, 1985).

Sociocultural theory. This concept appears to account for a variety of ways in which adjustment to old age may occur. Social context and expectations play a part in the morale and behavior of aging persons. Society sets the norms. The individual too plays a part; e.g., how one views the role of aging and one's resources, skills and perspectives influence the effectiveness of coping with physical, career, family, and motivational changes of aging. Research indicates that the processes and perspectives on aging are strongly influenced by social class, personality, life experiences, and personal orientations (Brubaker, 1985).

Inadequacy of theories. Even cursory examination of the major theories of aging reveals incompatibilites and areas of void. For these reasons we ought to be more concerned about how to slow down physical decline and how to enhance psychological and social dimensions of later adulthood. As such perspectives are gained there will be a simultaneous gain in our knowledge about the enhancement of life and living in all the age-stages.

PRINCIPLES OF DEVELOPMENT

In this introductory chapter a variety of psychological viewpoints on development were discussed. Next some of the ways in which age-stages are unique and different were considered. We now present an approach to unifying the diversity of views and uniqueness of age-stages by citing some principles of development. These principles transcend diverse views and apply to all age-stages.

Development Is Interactive

Growth is a process of interaction between the organism and its environment. There are *three* (not just two) facets of this principle: organism, environment, and interaction. Development throughout life consists in what one *is* at the time of birth (hereditary potential), what one *has* (environment), and what one *does* (reaction to potential and opportunity).

Multiple Causation

The factors which contribute to behavior and personality are numerous and varied. It is a human trait (and a generally effective one) to look for the easy way, the simple solution, the direct path. However, the easy way is not always most effective. It sometimes leads to clichés (e.g., Spare the rod and spoil the child; Easy does it; or, You can't keep a good man down) and overgeneralizations. It is necessary to know that principles are tendencies, not laws or inevitabilities.

Initial Rapidity

Development is most rapid in the early stages. The fertilized egg grows at an astonishing rate in the nine months of intrauterine life. Infants triple their birth weight in the first year and are half as tall, generally, as they ever will be by age two. The major part of one's cognitive potential has evolved by adolescence. Moral character is well determined by age six. There are in this principle numerous implications for parents, teachers, and all citizens who have taken substantial steps toward developing generativity.

Differentiation and Integration

Development consists of simultaneous and compatible processes of differentiation and integration. Differentiation is clearly illustrated in embryonic growth. As the fertilized egg develops, the head appears first, then the arm buds are differentiated into arms, fingers, and fingernails. So too are the legs and internal organs differentiated in the fetal stage. Integration is illustrated in postnatal life when the arms make only initial sweeping, thrashing movements and then are integrated into behavior such that a one-year-old can pick up a raisin or a pea and quickly transfer it to the mouth.

In the statement of the principle, the word development could be preceded by the word infant, child, adolescent, or adult. It could be preceded by the modifiers skill (development), cognitive, emotional, or personality. For example, we might cite linguistic development in children as they differentiate from the inclusive word "doggie" the specifics of dog, cat, horse, cow, rabbit, etc. Then as one develops, sentences increase from two or three words to longer ones in which the six-year-old says, "Dad, I'm going over to Jimmie's house to watch TV" to the still longer sentences used by teachers and politicians to express differentiated and integrated ideas. Not only are precise words used to differentiate and clarify the idea precisely but inflection, speed of utterance, voice volume, and gestures are integrated into the presentation. Another illustration might be the skill of playing a piano selection by the composer Mozart to the acclaim of music critics. The skill of a golfer who hits a number 6 iron from the 150-yard marker to within a foot or two of the cup to score a birdie also illustrates differentiation and integration.

Drive for Homeostatic Balance

The organism seeks conditions, or pursues activities, that relieve tensions and maintain equilibrium. Homeostasis is illustrated by the heart slowing or accelerating in response to temperature changes and activity. The endocrine glands produce hormones that prepare for activity or hormones that counteract chemical deficits or excesses in blood chemistry. In spite of living in the frigid or torrid zones and in spite of wide differences in diet, the body temperature remains normally at about 98.6 degrees Fahrenheit, and the blood composition does not vary significantly. These processes of automatic adjustments are known as homeostasis. Speculations have been made about the implications of body chemistry for personality, intelligence, and behavior.

Psychological adjustive mechanisms (fear, anger, love, rationalization, projection, psychosomatic illness, etc.) do for the emotional aspects of living what the endocrines do for the physical. Moreover, adjustive mechanisms and endocrines work together—sometimes in harmony and sometimes in contradiction. However, the goal in stress situations is homeostasis.

There are many kinds of behavior that are expansive and exploratory that do not seem to fit into the concept of homeostasis. Therefore, the drive for homeostatic balance needs to be supplemented with another hypothesis.

Drive for Heterostatic Activity

The normal individual has a desire to explore, to learn, to satisfy curiosity, and to "become." This thrust, a counterpart to homeostasis, is a reaction against the idea that all motivations stem from deficiency. The principle recognizes the potency of curiosity as a motivation in young mammals. It questions the idea that humans are simply seeking to escape from tension, change, anxiety—to reach a state of rest or balance. The heterostatic drive to grow, to be independent, to "do it myself," to learn, to exert influence, does not have to be taught. Young and healthy persons of all ages have this heterostatic drive in abundance. However, there are many things (demand for conformity, obedience, formal school requirements, authoritarianism, too heavy burdens, etc.) in our culture that inhibit and discourage its manifestation, exercise, and enhancement.

Problem finding is an example of the heterostatic drive. It is now perceived as a prominent characteristic of those dreamers, questioners, and off-beat thinkers who are identified as being outstandingly creative.

The practical implications of the principle of heterostasis are matters of understanding rather than plans for action. We can now understand why children climb trees and adults climb mountains. We can understand why boredom is as big a threat to ego security as is continuous uncertainty.

Development Is Continuous

Relative rates of growth in trait development vary but are nevertheless continuous and gradual. This principle is especially important as a counterbalance to becoming overly enthusiastic about the principle of initial rapidity. There are times of relatively rapid growth, or critical periods for certain aspects of development. For example, the first two years of life are considered to be critical in the establishment of human bonds; adolescents have to relate to the other sex as another step toward intimacy; and adults must (in order to achieve mental health) move toward intergenerational concern. In short, now is the time. It is not too late after the period of initial rapidity of development for growth to occur.

Training Influences Development

The effect of training varies with the individual's level of maturation. Children cannot be taught to walk or talk until they are ready. Youngsters typically do not profit from reading instruction until, among other things, they have reached a mental age of about 6½ years. Boy-girl relationships are best learned during adolescence; and, if such relationships are missed at that time the chances of gaining intimacy are markedly diminished. Havighurst's (1972) developmental tasks, Erikson's (1980) ages of man, Kohlberg's (1983) moral stages, and Piaget's (1970; 1981) cognitive levels of development are all based on this postulation of relating instruction or guidance to status of maturation.

Development Is Sequential

Growth is an orderly, systematic, sequential process. This means that human development occurs in steps and stages that follow one another in a reliable and rather predictable manner. For example, the emotional outbursts of infantile rage are replaced, step-by-step, by the more specific and better controlled and directed emotional expressions as children grow older. Deviations from this progressive, sequential pattern of development—be it in walking, talking, playing, socializing, formal learning—is a symptom that something is wrong.

The need for remedial intervention is indicated. The sequential nature of personality development may be illustrated by a theory of play. The first stage is solitary play in which children play by themselves. Later they play near but seldom with other children—dominion and parallel play. Interactive play goes through the levels of random and casual interaction and finally into cooperative and competitive team play.

In a roughly similar manner humans move from highly egocentric concern to caring about others. Infants typically are tolerated when they express their desire for immediate satisfaction of wanting to be fed, held, cuddled, or relieved of pain. Older children and certainly adults are not so readily tolerated if they have not taken some of the sequential steps toward the more mature manifestations of development.

Such manifestations are expressed in various ways. The highest level of personality development is, according to Jersild (1968) compassion; Maslow (1970) sees self-actualization and aesthetic appreciation as manifestations of high levels of development; Erikson (1980) suggests generativity and ego integration are high on the level of human development. On the moral plane, Kohlberg (1983) postulates acting on universal ethical principles as the model of humanity. No matter how the high level of development is expressed, or in whatever area, it must be preceded by orderly sequential experiences and stages.

Correlation and Compensation

Subaspects of development within an individual tend to be positively correlated rather than for weaknesses to be compensated for by strengths. This is an important principle because popular misconceptions which deny it are widespread. For example, the cliché "Strong back, weak mind" is generally incorrect and injurious. Terman and Oden (1947) years ago found and reported that mentally gifted children were age for age, and year after year, stronger, bigger, healthier, had better vision, more curiosity, and stable mental health than "normal" children. Nevertheless, we continue to express the compensation idea.

The major implication of the correlation principle is the need for understanding on the part of parents, teachers, and peers. In the first place, the principle states a tendency or probability, not an infallible law. There are times when a mentally gifted child is emotionally unstable or has sensory defects and is delicate. There are academically slow pupils who have remarkable talents in music, athletics, or dramatics. When these talents are observed they should, of course, be capitalized upon as sources for giving the child the satisfactions of recognition and achievement. But, equally as pertinent, the temptation to try to create a compensatory skill or talent should be avoided when there is no latent capacity.

Chapter 2

MAINTAINING HEALTH AND WELLNESS

Physical growth is an integral part of human development. When one considers the marvel of conception, heredity, and prenatal growth, the process is nothing short of one miracle after another. Just the sheer growth of the fertilized ovum—a bit of matter no bigger than a period—to a seven or eight-pound baby is an awesome phenomenon. That ovum has become a baby complete with toes, fingers, eyes, ears, and brain more complex than the largest and most capable computer. The nine-month wonder of internal, automatic growth is no greater than the marvel of what the newborn baby can do and become by the time adulthood and maturity is achieved some two to six decades later.

In this chapter we will (1) consider some of the physical processes of development, particularly those that are amenable to external influence and intervention; (2) identify obstacles to good health during each of the age stages; and (3) describe the conditions and personal choices which promote good health and wellness.

WELLNESS: AN ISSUE OF PERVASIVE RELEVANCE

No matter what description of human development is taken (the topical approach we are using or the more conventional age-stage approach) some topics do not readily fit a sequential discussion. For example, health and wellness is a concern of all ages and its ramifications are emotional, social, and cognitive as well as physical.

Health may be defined as freedom from physical disease or pain. Wellness relates to feeling good, the zestful motivation to make optimum use of one's potential for strength, agility, and the handling of toys, tools, persons, or objects. Wellness can coexist in the presence of defect or disease. Such matters as exercise, diet, medical care, mothering, avoidance of substance abuse, and hygiene are involved in both health and wellness. The French term, elan vital, embodies some of this concept. It means, in part, the zest for living and growing. Sometimes it is referred to as heterostasis—the urge to grow, reach out, be active.

Infants are helpless, so their health and wellness needs must be supplied by knowledgeable affectionate caretakers. The need for concern by others may be less apparent for wellness in children, adolescents, and adults because of cultural pressures and practices. But the need for cross-age concern functions throughout life. One manifestation of the needed concern is instructional and the other involves psychological support. To illustrate, an early medical discovery was that hygiene and cleanliness nurtured health. Centuries before antiseptic operating rooms, or surgical instruments, and before germs had been recognized, doctors knew about the need for hygiene in the prevention and cure of disease. Yet today in developing countries, in city and rural slums, or even in middle and upper class homes, basic cleanliness is sometimes neglected—washing hands after handling pets or using the toilet. It is in this simple area of application of what is known that adults can help children and adolescents. The concept is illustrated by neighborhood clinics for parents-to-be.

Students should be encouraged to practice the health and wellness principles taught to them in elementary and high school. Contagious diseases, AIDS, herpes, colds, and influenza are still health problems in the 1980s. Specific recommendations for staying well include: eat wholesome meals; do not smoke; use alcohol in moderation; learn to cope with stress (i.e., daily intervals of relaxation); exercise regularly; know your sex partner; and sustain warm, friendly relationships (there is a positive correlation between loneliness and susceptibility to infection). Even old persons need to be reminded or encouraged (perhaps by children or teenagers) to eat properly, take their medications, and to exercise.

In addition to hygiene, ancient medics recognized the wholistic nature of health and wellness. They knew, as Benson and Proctor (1984) and Cousins (1984) are emphasizing the inseparable (wholistic) nature of mind, emotions, and body. What we believe and feel, says Cousins (adjunct professor of psychiatry and biobehavior science at the University of California at Los Angeles), has an effect on health. Benson, a physician at Harvard, suggests that religion and medicine cover the same considerations—life, death, and well-being—but from different perspectives. It is time, he says, to combine again (as did the ancients) faith, belief, and trust with medicine and health practices.

Benson states that the popular disease of the twentieth century is hypertension. Anxiety, stress, and time-pressures contribute substantially to heart attacks, atherosclerosis, and strokes among adults of all ages. There are ways to reduce the impact of stress. Among other things, working fathers and mothers are being asked to consider fewer hours on the job and more hours with their children. In any event, plan to spend time off with sons and daughters of all ages. Dividends accrue not only to the children but the parents and their associates.

Many physicians, along with Cousins and Benson emphasize that positive emotions can be therapeutic. Doctors use the patient's hope as an essential aspect of therapy and recovery. Witch doctors and medicine men in "primitive" cultures use faith and hope in the successful treatment of their patients. Cousins reports on a cardiologist who asserts that no doctor should tell a heart patient that his case is hopeless, because learned specialists simply do not know all there is to know about the role of belief and emotions. "Scared to death" and "died of a broken heart" are not just figures of speech. Emotions can be fatal (Goleman, 1984). Many doctors say they can treat the disease but the cure depends on the role patients will permit, or force, their inner resources to perform. A well-known example is the influenza epidemic of 1918 which took one or more lives in most families in the United States and Europe. Often the mother, grandmother, or oldest daughter on whom all the other family members depended—despite close and continuous contact with the disease—escaped. That person simply could not afford to die. Such reports—that science tends to ignore because they are not replicable—tell us of fathers, mothers, siblings, or others who have been able to instill the elan vital that can revive seriously ill persons.

Health and wellness can be used to illustrate the principles of growth cited in Chapter 1. Especially prominent in this context is the principle that forces are interactive (mind, body, and emotions); that causes are multiple (self, physical and social milieu); and that health and wellness are a continuous living concern. The effect of learning (applying what is known) determines whether or not our study is worthwhile.

Prenatal Development

Critical Periods

A critical period is a span of time, typically short, during the course of development in which the organism has an enhanced susceptibility or responsiveness to given conditions or experiences. For example, following the moment of conception (when one male sperm penetrates the female ovum) there occurs an eight-week span of development called the period of the ovum and embryo. During this time toes emerge on the leg stubs and fingers on the arms. Eyes, mouth, an ears appear on the embryo. If, during this sensitive period, certain trauma occur within the intrauterine environment, the baby to be born a few months later may have no toes or fingers. So, the first eight weeks of prenatal life is a critical period for certain aspects of development.

The first highly publicized example of how drugs can cause birth defects occurred in the 1960s. Thalidomide, a tranquilizer, was taken rather freely by pregnant women in Western Europe to relieve the nausea of morning sickness. Use of the drug was frequently accompanied by the birth of physically abnormal babies, e.g., infants had no legs, one arm, or an arm stub with no fingers. By the time the link was recognized, more than 10,000 children had become victims. Thalidomide taken after a certain period of pregnancy has no such effect (Apgar & Beck, 1972). At about 20-25 weeks of pregnancy fingers are appearing on the fetus so it is obvious that the drug could no longer result in an armless child. The critical period has passed.

Another tragic example involves Benedictin which for almost 30 years was the only government approved prescription drug for morning sickness in early pregnancy. From the time this drug went on sale in 1956 until it was taken off the market in 1983, an estimated 33 million women worldwide took it. A class action suit filed by many women blamed Benedictin for their children's congenital deformities involving hands, arms, feet, legs, and other muscular-skeletal problems.

Avoidable Prenatal Hazards

There are other threats to normal fetal development such as sexually transmitted diseases (STD). Vitamin deficiency, tranquilizers, and aspirin may have negative fetal influence. Obstetricians have recently cautioned against the use of Lithium, Tetracycline, and the popular acne drug, Accutane during pregnancy (Winick, 1985). Expectant mothers who had used Accutane were 26 times as likely to have a child with serious birth defects. Some of the factors which cause birth defects for 250,000 babies each year are subject to external control. Heredity itself is subject to control as genetic knowledge grows at the rate of doubling every two years. Three thousand human genetic diseases have been identified and genetic screening, prenatal diagnosis, and fetal therapy have become practical realities (Harper, 1985).

Each organ of the body is particularly sensitive (another name for the critical period) to trauma during the time when the organ is growing most rapidly. For example, rubella (German measles) often has the effect of causing hearing problems, retardation, or heart and eye defects if the mother suffers the disease during the first three months of pregnancy. The placenta is no barrier against the virus. Girls should be innoculated against measles to prevent this particular tragedy. Despite the best efforts of health organizations, about one in seven women of childbearing age remain unvaccinated and thus susceptible to rubella (Brewer & Brewer, 1985).

Not so specific in regard to timing is the impact of sexually transmitted diseases, the effects of which may be passed from mothers to their children. About a quarter of the babies born to syphilitic mothers have physical or mental abnormalities. The incidence of spontaneous abortion or stillborn babies is higher than in births to healthy mothers. Children born with gonorrhea, if untreated, are likely to suffer heart disease, blindness, and arthritis. Similar damage can be expected in cases of chlamydia, a sexually transmitted disease which is twice as common as gonorrhea. There is also growing concern about the prospect of brain damage, retardation, and death for children of mothers with herpes. A pregnant woman with active genital herpes (HSV-2) can infect her baby with the virus during delivery. To prevent this danger, a mother may have her baby by caesarean which avoids going through the infected birth canal (Holmes, Quinn, Corey & Cates, 1985).

Physicians working in reproductive medicine state that there is no drug (including nicotine and aspirin) used by pregnant women that does not reach and affect the fetus. Drugs are particularly hazardous during the first 12 weeks of pregnancy and they continue to affect development postnatally. Some newborns show the anguish of drug withdrawal, while others may merely be handicapped by their small size, irritability, prematurity, and diminished responsiveness. If a mother wants a healthy, average weight, well developed, serene baby she should not smoke, drink alcohol, or take any unprescribed drugs.

It is important for pregnant women to understand that everything they eat or drink gets to the fetus through the placenta, a thick wall of tissue through which the baby's nourishment comes from the mother's bloodstream. While certain substances have difficulty getting past the placenta, others such as alcohol and many medications pass through quickly. Some accumulate in greater concentration in the fetus than in the mother. One reason babies are so significantly affected by drugs is that they lack the liver enzymes, which only function after birth, to break down drugs. Consequently, the fetus is incapable of protecting itself.

Teachers, nurses, social workers, and parents need to take a position on drug usage if they are genuinely concerned about the welfare and wellness of babies and children. Smoking and use of alcohol by pregnant women is widely condemned by medical research. Smoking during pregnancy exerts a retarding influence on fetal growth. There is decreased average birth weight (less than 5.5 pounds) and increased incidence of premature birth among smoking mothers. Unlike Benedictin and rubella which have their most serious consequences in the first trimester of pregnancy, smoking has its most damaging impact in the last trimester (Bennett, 1985a). Much of what can be said of developmental influences must be stated in terms of probabilities—not certainties. Apgar and Beck (1972) contend that "no effort is too great to increase the chances that a baby will be born without handicap."

Prenatal Diagnoses

Down's syndrome and other abnormalities in the unborn can be detected by a technique known as amniocentesis. A needle is inserted into the uterus and some amniotic fluid is removed. The disadvantage of amniocentesis is that it cannot be used until the second trimester (about 16 weeks) because there is not enough fluid until then. It is also necessary to wait three or four additional weeks for the test results. Then, if a baby is identified as defective, the parent choice is to deliver or have a saline abortion. A new test called chorionic villus involves taking a sample of the placenta. The advantage is that results are obtained during the first trimester.

Another medical technique, one that has revolutionized obstetrical care for high-risk pregnancies, is called ultrasound imaging. Using a device called a transducer, high pitched sound waves, inaudible to the human ear, are sent through the amniotic fluid. When the sound waves bounce off fetal structure, they are collected as two-dimensional pictures on a video screen. This method, an adaption of sonar used in World War II to detect submarines, allows the physician to see the unborn patient. Detailed measurements of head, heart valve action, exact age, and position of the fetus can be determined. It is also possible to identify certain birth defects like spina bifida (defect of the spinal column) or hydrocephalus (accumulation of fluid around the brain). Approximately 35 percent of American expectant women are given an ultra sound test at least once during their pregnancy (Winick, 1985).

BIRTH

Delivery Preparedness

Natural childbirth is steadily gaining acceptance and emphasizes that normally the use of drugs to reduce pain is unwise and can be harmful. The term natural childbirth does not mean a return to primitive, do-it-yourself practices. Prepared childbirth should begin as soon as pregnancy is confirmed. Lectures and movies are used as doctors or nurses describe the physical processes and cautions. Questions are answered by understanding professionals who know that women need support during pregnancy and labor. In the United States many birth preparation classes are based on the Lamaze method which emphasizes breathing patterns. Exercises and practice to strengthen and relax muscles are part of a regular routine (Lamaze, 1984).

Prepared childbirth may, but not necessarily, involve the use of drugs and instruments. Because some women are temperamentally unsuited to participate in natural childbirth, it may be appropriate to administer light doses of drugs. Natural childbirth recognizes individual differences—what is right for one mother is not necessarily right for another (Spivak, 1985).

The emotional and physical cannot be separated except for convenience in academic presentation. Preparation for successful childbirth begins with the development of good mental health habits and pursuit of wellness in one's own childhood. Some headaches, nausea, weakness, and discomfort during pregnancy can be attributed to fear, marital disharmony, and tensions existing between mother-to-be and other family members. The tone of the relationship with her spouse conditions her psychological and physiological state—and that of the fetus. Marital relations after the birth are crucial factors in the psychological development of the child.

Birth Trauma and Mortality

Part of the folklore of childbirth is the belief that neonates (the first 28 days after birth) experience trauma when thrust into the cold, cruel world from the warm, soft, nourishing environment of the mother's womb. Otto Rank (1973) taught that the trauma of birth lasted a lifetime; Sigmund Freud (1962) also postulated that anxieties generated at birth might become permanent personality traits. It seems more sensible to believe that when a fetus achieves a weight of about 7 pounds and has the normal maturity expected in full term babies, the traumatic thing is not to be born. When one is ready for more freedom and space, the truly traumatic thing would be to remain where food, oxygen, and room for movement was restricted. Trauma can, of course, occur in abnormal cases. As already indicated, mother's smoking may make labor prolonged and difficult. Abnormality of pelvic structure, breech presentation, or delayed birth resulting in overlarge infants may make the use of instruments or caesarean birth advisable—with an increased likelihood of trauma. At the present time children born in the United States are more likely to survive and develop into healthy adults than were children born several decades ago.

MAINTAINING HEALTH AND WELLNESS   57

INFANCY

Assessment of Growth

Infancy is defined as that phase of postnatal life during which the baby is totally dependent on caretakers (under age two). Being unable to walk or talk, others must see to it that the infant's needs are met. Despite the infant's helplessness, some remarkable changes take place.

Before 1960 height-weight tables were typically cited in child development books. Because there were so many exceptions to the average which caused parents to worry, the presentation of average tables was abandoned. Each child has its own unique rate and normal variations are wide. Concern has shifted from size to the sensitivity and responsiveness of infants and children. For example, the Apgar Scoring Chart evaluates the neonate's heart rate, breathing, muscle tone, irritability, and skin coloration. The Bayley Scales of Infant Development and the Denver Developmental Screening Test are used to assess physical development and motor skills.

Hunger and Nutrition

The problems of hunger and starvation are not limited to places where drought temporarily disrupts farming; where there is overpopulation or where distribution problems contribute to farmers' going broke. More than 100 countries consume a greater amount of food than they produce. Consequently, the fight against hunger calls for humanistic collaboration among governments and private agencies on an unprecedented scale. It is important to recognize that malnutrition also exists in our own affluent society. Concisely, the data are:

> Presently one-fourth of all families in this country have incomes below the poverty line.
>
> Forty percent of all children under five years of age in developing countries suffer from protein malnutrition.
>
> Every day 35,000 people worldwide die because of

starvation—two thirds of them are children (Williams & Kornblum, 1985).

Because the infant brain is in its last stages of most rapid development—a sensitive period—its nurture is of prime importance. Despite this importance, study of brain growth until only recently has been almost totally neglected in the study of human development. But at a cost. Some scholars estimate that as much as 90 percent or more of our cognitive potential is unrealized. However, there is progress. In addition to study of environmental milieu, the brain chemistry involved in mood, emotion, and thinking is a focus of current research (Whalen & Simon, 1984).

It is realized that physical nourishment is one factor in brain development and function. To illustrate, a diet rich in carbohydrates increases the rate at which the brain synthesizes the neurotransmitter serotonin—which facilitates the transmission of signals between the neurones within the brain. Animals deprived of serotonin become insomniacs. Protein rich diets raise the level of amino acids competitive to serotonin and hence lower the level of neurotransmitter action. The finding that food affects brain and brain affects food consumption cannot be dismissed as mere coincidence. Study of these transmitter systems has already provided significant clues to the chemical mechanisms involved in learning, memory, sleep, and mood (Lynch, McGaugh, & Weinberg, 1984).

We are not ready to say that "You are what you eat." However, it is certain that diet has considerable and controllable effect on brain development. Dietary deficiencies in the prenatal and postnatal periods may cause children of lower socioeconomic status to get a slow start in mental development and suffer that handicap irrevocably. Mental growth is most variable and vulnerable at the time when physical growth is most rapid; i.e., prior to school age (Kim, 1985).

EARLY CHILDHOOD

Children's Needs

As boys and girls move from infancy to childhood their needs become more personal and social. They need to grow, acquire skills, and to be accepted in terms of their individuality. Like infants, they need love, but their cognitive progression arouses a need to belong.

Like adolescents and adults, they need to achieve; but this need is not so prominent as the need for love and acceptance—which is supplied by others rather than from their own strength. In order to meet these expanding needs, children require a secure environment.

At the beginning of the century disease was the major killer of children. Diphtheria, tuberculosis, pneumonia, and meningitis were the leading causes. Currently, accidents—automobile crashes (failure to wear seat-belts), fires, falls and drowning—cause the most deaths among children (Allman, 1985). Obviously children need to be protected from disease, and such things as knives, fire, vicious dogs, and cars.

Boys and girls also need the safety of parents who are dependable. They need freedom from fear, anxiety, and chaos. This need is met by a caretaker who is perceived as being powerful and reliable. Parents who do not consider themselves in charge and responsible cause children to be confused. Youngsters need to perceive stability in their caretakers. Parents who frequently quarrel and present a scene of instability undermine the safety needs of their children.

During the past few years parents have found it necessary to caution their young children against interacting with strangers. On the one hand, parents want to encourage a sense of trust but they are also aware that kidnapping and sexual exploitation are on the rise. In order to protect boys and girls, a National Center for Missing and Exploited Children has been established; television stations carry child watch reports detailing the physical characteristics of missing youngsters; local police departments encourage parents to have their child fingerprinted; dentists implant microdiscs in children's teeth carrying identification numbers filed in computer registries; community agencies advise parents to make videotapes of their child and keep up-to-date photographs. The safety of children has become a public mission with good results. During a single year over 2600 children from 14 months of age to 17 years were found by the National Center and its collaborating agencies. Most of the missing (2000) had run away from home; some (400) had been snatched by estranged parents; others (60) were kidnapped by strangers. All but 30 of the children were found alive (Meddis, 1985).

The need for adult affection and orderly routine continues as children enter daycare and preschool programs. Some intergenerational projects are attempting to help meet these needs. In California, the first state to pass an intergenerational childcare act, centers for two- to six-year-olds of working parents include elderly helpers on the staff. Previous to their part-time minimum wage assignment, the 55- to 75-year-old helpers receive training in child development which consists of observation, reading and discussion with teachers as well as parents. Another intergenerational effort is the foster grandparent program currently operating in 37 states. Here the older participants work to ease the caretaking burden for one-parent families, mothers and fathers of retarded children, school-attending teenage mothers, and child abusive couples. Similarly, in Washington, D.C., the Family Friends Project matches older nonprofessional volunteers with chronically ill or disabled children and their families. The trained elders come to the child's home to provide assistance (Thorp, 1985).

Going to school at an early age involves some health risks. The once popular belief that children should be exposed to chicken pox, mumps, and measles in order to get it over with is unjustified. Even minor illnesses temporarily retard growth in preschool children. These illnesses may leave tiny scars on the ends of growing bones that are discernible by X ray. One of the major hazards of preschool and daycare centers is the high frequency in which children catch colds, flu, and other minor diseases. Studies have shown that members of this age group put a hand or object into their mouths an average of once every three minutes (Howell, 1985).

Child Abuse

One evidence that some children's safety needs are not met is the phenomenon of child abuse. Since Congress passed the Child Abuse Prevention and Treatment Act in 1974, the number of reported cases has risen dramatically. Doctors, nurses, social workers, and teachers who suspect abuse are required by law to report their suspicions. Approximately one million children are reported abused each year (Meir, 1985).

Boys and girls who are abused tend later to become abusive parents, abusive spouses, and mistreat their parents. The correlates (but not the causes) of child abuse are low income, broken marriages, psychiatric instability of parents, and inadequacy of parental socialization. Whatever the contributing factors are in a given case, the prospect of next generation child battering remains. The dismal prospect is avoidable, though there is no simple solution. Education, economic conditions, cultural tradition (parents are rulers in their own homes and can be tyrants over their children) and mental health are involved. Battering parents can be reoriented or rehabilitated. As these parents are helped they recruit to therapeutic circles other people who are reluctant to seek and accept help.

MIDDLE CHILDHOOD

Growth and Health

The physical changes of middle childhood are ones of proportion more than increasing size. The rate of growth is decelerating. The trunk grows longer and slimmer, the chest becomes broader and does not have the rounded appearance of babyhood. By eight years of age the arms and legs are nearly 50 percent longer than they were at the age of two, yet overall height has increased by only 25 percent. Along with the eruption of permanent teeth, the lower part of the face becomes more prominent. The head, once large for the size of the body, now more closely approximates adult proportions.

Middle childhood is a healthy period. The death rate during this period is the lowest of any comparable span of years. The kinds of diseases that used to account for most cases of middle childhood hospitalization—rheumatic fever, rheumatic heart disease, osteomyelitis, pneumonia, streptococci infections, meningitis, polio—have declined markedly. Today accidents are the largest single cause of deaths during middle childhood—responsible for 50 percent of boys' deaths and 25 percent of girls' deaths (Allman, 1985).

Physical Fitness

Although middle childhood is a time of relatively good health, judged in terms of mortality rates, it is not a period of good physical fitness. Most public school students do not get enough aerobic activity to keep their heart and lungs fit. In a nationwide assessment over 85 percent of the participating children could neither stand with knees straight and touch the floor or do a single situp with feet held to the floor. This low fitness score is attributed to such things as:

> The large amounts of time preschool children spend watching television, thus curtailing their opportunity for physical play.

The taxpayers revolt against financial support for schools. They want better teachers and a return to the basics but curtailment of such "frills" as physical education.

Insufficient time in the school schedule is spent on physical exercise.

Parents who are tennis, handball, or jogging buffs permit children to be spectators, or simply leave them home to engage in sedentary activities.

School programs promote interscholastic athletics which develop those who are gifted or precocious athletically and ignore the rest (Harris & Gurin, 1985).

The relationship of physical health, vigor, and fitness to other phases of development is illustrated in studies of boys' size, athletic prowess, maturity, social adjustment, and leadership. It has been found that boys who are large and relatively strong for their age make better social adjustments than those who cannot participate so successfully in play and games. Relationships between body build and personality, although well established, may at least in part be due to the attitudes of caretakers and their acceptance of the child. Many comparatively small children are well-adjusted, self-confident, and socially competent. A more extreme example reported by Halverson and Victor (1976) is that handicapped children may or may not develop the behavior traits that typically are attributed to specific defects. Health and physique are influences on personality, not determinants.

Obesity

Obesity is a self-perpetuating handicap. Obese children tend to become obese adults. Children (and older persons) often overeat because they are lonesome or bored. Then their obesity contributes to their being ignored by peers and left out of activities and games. Later they eat more to compensate for the initial lack of socialization.

There is uncertainty about whether fatness is inherited or is socially conditioned. Obesity represents 15 to 20 percent of significant medical pathology and begins in childhood with the number of fat cells becoming fixed at about age ten. Parents should know that obesity in childhood merits immediate concern. Mayer (1975) found that 7 percent of children of normal weight parents were obese; but 80 percent of those with two obese parents were overweight. This does not resolve the issue of heredity versus social conditioning. There is also the matter of activity. Obese girls play tennis or other games and move only about 60 percent as much as normal weight girls in the same game. Obese girls could be helped to reduce by a regular program of exercise without cutting the amount of food eaten (LeBow, 1983).

The implication of obesity studies are (1) avoid overfeeding of infants because this activates the growth of fat cells; (2) attack obesity indirectly by giving attention to social adjustment; (3) attack obesity directly by exercise programs; and (4) obesity in childhood persists in succeeding years (Turner & Aschner, 1985).

LATER CHILDHOOD

Physical Growth

The growth spurt of later childhood is intrinsically determined by changes in glandular activity. Specifically, after a dormant period from birth on, the glands begin to secrete sex hormones that stimulate growth of the reproductive organs. The pituitary gains in size and increases the secretion of gonadotrophic hormones. The peak of the spurt occurs in about the middle of the short span of time during which the gonads undergo marked changes in size, pubic hair appears, and the breasts of girls enlarge. This accelerated growth period is called puberty and refers to the time and processes involved in the marked enlarging of the reproductive organs.

Growth velocity decreases after about age eleven or twelve—a little earlier in girls than in boys. With due consideration for wide individual differences, there is a rhythm of growth that occurs in somewhat the following pattern (Reese, 1983):

1. rapid growth from birth to seven or eight years, with continued deceleration;
2. slower but still steady growth from seven or eight to about ten years;
3. an accelerated growth period, lasting about two years,

begins at about age ten to twelve, depending on the individual's sex and internal time-clock;
4. slowing of growth rates begins at about age 14 or 15, with growth virtually ending at about 19 or 20;
5. subcutaneous fat, especially noticeable in the limbs (the fat has been present since infancy but leveled off during early and middle childhood) accelerates during later childhood. At adolescence this fat decreases, more for boys than girls, but there is no loss in total weight.

Girls tend to be about a year advanced, physically, over boys of the same age during the final part of preadolescence. However, boys in the next period of development—the early adolescent years—make up the difference and will again, on the average, be taller and heavier than their female agemates. Actually, despite the average boy-girl differences, the developmental contrasts in size within the sex are greater than average differences between the sexes.

Brain size in preadolescence has achieved virtually adult status, but intelligence will continue to grow. Eye development at this period has reached mature size and function. There is a great improvement in muscular coordination and manual efficiency. There is a marked increase in strength, and boys particularly (due in part to cultural expectations) want to demonstrate their prowess. There is an increased tolerance for the expenditure of energy. Most youngsters at this age level seem to have limitless energy and often worry parents who think they get insufficient rest.

Vision and Hearing

One in every four school age children are affected by eye problems. There is a wide range of visual handicap, from a need for glasses in certain situations to total loss of vision. Even blindness is not a unitary concept. There are the legally blind, the medically blind, and the occupationally blind, as well as the partially sighted and visually impaired. As a rule, a person is considered blind if the better eye tests no better than 20/200 after correction and visually handicapped if the better eye tests between 20/70 and 20/200 after correction.

Annual screening of vision should take place at school because some children otherwise go unchecked. Teachers ought to be aware of symptoms of visual handicap and of the need for careful, immediate, and professional referral if symptoms appear. More specifically, educators should watch for squinting, tilting the head, rolling the eyes, inattention to visual objects (books, pictures), avoiding work requiring close vision, and getting the head down close to paperwork. It is estimated that nearly 25 percent of the American population is nearsighted (Kolata, 1985). Students known to have visual defects can be given the best table or chair for seeing the chalkboard or center of activity, with light coming from in back of the child. Tasks calling for prolonged use of vision can be interrupted periodically.

About ten percent of the school age population have hearing problems. Hearing handicap is not an either/or phenomenon but a matter of degree. Persons may adequately hear low-pitched voices but miss some sounds, especially the high-pitched ones (s, z, p), when spoken by those with high-pitched voices. Some may hear adequately in a room where there are no competing sounds—other voices, traffic, fans, music—or where echoes are minimized by room structure or accoustical tile ceilings. Out of these varied conditions come the common misconceptions "He can hear when he wants to" or "She can hear if she just pays attention." In order to maximize school achievement and social adjustment annual hearing checks should be conducted by the school nurse (Bennett, 1985).

Adolescence

Physical Characteristics of Adolescents

Before cautiously describing the physical characteristics of adolescents, it is necessary to reiterate that as persons grow older the greater their human differences become. There is no typical adolescent. Some normal boys are smaller than most girls and some normal boys, at age 15 or 16 are larger, by far, than their fathers. The average age of first menstruation (menarche) for American girls is 13; but normal girls may menstruate before age 12 and some not until 17.

Growth processes of children and adolescents are accelerating throughout the world. This is called the secular trend. Age for age in North America, Europe, and in other developed nations, teenagers are taller and heavier than their counterparts of 30 years ago. In the United States average boys grow to be an inch taller and ten pounds heavier than their fathers. Girls grow one-half inch taller and two pounds heavier than their mothers. Differences might be greater if it were not for the fact that puberty occurs earlier and growth ceases correspondingly early (Schuster & Ashburn, 1986).

Adolescent Glandular Development

Adolescence is a stage of maximum vitality. The most noteworthy physical changes that occur stem from altered endocrine functioning. The activity of the pituitary gland, which is primarily responsible for growth, is inhibited by the hormones from the gonads, and growth rates of bodily mass are sharply decelerated.

In the female, hormones from the ovaries stimulate development of the breasts, mammary glands, uterus, fallopian tubes, and vagina. These hormones also produce the secondary sex characteristics, including pubic and axillary hair growth and increased activity of the sweat glands. The hips tend to broaden, making room for the increased size of the internally located female sex organs. In general, there is a change from the angularity of late childhood to the roundness of adolescence.

In the male, the hormones from the testes stimulate growth of the prostate gland, seminal vesicles, and penis. The hormones also change the secondary characteristics of heavier hair texture, broadened shoulders, deeper voice, and, as in the female, growth of pubic and axilary hair and increased activity of the sweat glands. Muscles tend to strengthen and enlarge, and in many cases the plumpness of late childhood is replaced by firmness and symmetry of body build.

Typically, all these hormonal changes seem to have effects on behavior that are easily absorbed. If an adolescent experiences puberty at an unusually early or late age, he or she may be concerned about not fitting the developmental pace of age peers. But it is likely that the concern is compounded or reduced by parental treatment, peer relationships, and school achievement. In short, physical differences may become an excuse for but not the cause of emotional, cognitive, or athletic failure (Malatesta & Izard, 1984).

Accidents and Suicides

The teenage years are among the healthiest of the entire lifespan when judged in terms of mortality rates and causes of death. The major cause of death between ages 15 and 24 years of age is automobile accidents. Teenagers cause five times as many traffic fatalities as drivers aged 35 to 64 (Allman, 1985). The recent increasing suicide rate of adolescents is a growing concern. In the United States, approximately 6000 teenagers commit suicide annually. Girls try suicide about three times as frequently as boys. The shocking rise in the suicide rate can be in part attributed to shifting moral codes, parental instability, broken homes, swift vocational changes that make for uncertainty in job choices, and staggering array of curricular choices in school and college. All this makes for adolescent bewilderment and discouragement. This bewilderment is not restricted to teenagers. Teachers, counselors, and parents are also confused and can give little advice about these choices. Somehow we must all learn to feel more comfortable with continuous ambiguity.

Brent (1985) reports that teenagers who take their own lives are ten times more likely to be intoxicated or under the influence of drugs than their counterparts of twenty years ago. And, if they are drunk, suicidal teens are seven times more likely to rely on a gun as their means for self destruction. When a gun is chosen, as compared to other methods, the result is almost always death.

The line between accidents and suicide is often difficult to draw; it is thought that some accidental deaths are actually suicides. Suicidal people are often susceptible to depression. Usually they make attempts to describe their feelings to others but they are unable to make meaningful contact. There is a dangerous myth that those who talk about suicide do no commit suicide. Suicidal persons do talk about it but they need someone who will listen. Over 80 percent of the teenagers who have thought about suicide have communicated such intentions to parents or other key persons.

Disrupted communication between parents, teachers, and counselors and coping with rapid change as contributors to suicide show that the dilemmas of teenagers cannot be separated from the problem of other generations. The futility of studying the development of young persons apart from the adjustments being made by young parents, middle-aged teachers, and employers, and the lives of older persons is again demonstrated. The one antidote to suicide we do have demonstrates the merit of a cross-age approach. "Crisis Hotlines" show the potential of communication. The hotline provides potential victims with someone to talk to. These hotlines may be manned by persons of various ages who do not give advice. They refer the call to counselors, psychologists, psychiatrists, or clergy who are on the response team. The focus of these professionals is to replace the hopelessness with hope.

Accidents and suicides are only part of the health considerations during adolescence. Some persons undermine their health by smoking, drug usage, improper diet, and inappropriate rest and exercise. There are others, probably more numerous, but not so well publicized, who use better health practices than do adults. We suggest that teacher and parental pronouncements about drugs, smoking, and diet have less beneficial impact than exemplary adult behavior. This is true also in regard to tennis, handball, racketball, jogging, and dietary practices. Adult interest plus provision of opportunity to communicate and example can do much to provide a better milieu for health and wellness in adolescence.

Because we believe that adolescence is a social and cultural phenomenon rather than a physiological one, we endorse a social approach. Specifically, adolescents want and need to feel responsible and autonomous. Group discussions should be encouraged, allowing them to understand the risks of drug usage, smoking, drinking, and the importance of health maintenance. In those instances where autonomy and responsibility have been encouraged through discussion in the classroom and at home the usual results have been that the decision for autonomy was wise. The role of teachers or counselors in such groups is that of facilitator rather than expert or authority.

Early Adulthood

Physical Characteristics

The physical changes occurring in early adulthood are not so obvious as they were for prior age-stages. In certain athletic activities early adulthood sees the peaking of some skills but in other sports disinterest has already begun. This is the physical prime time of life, but, we must add our postulation that the intellectual prime of life is yet to come. Visual speed has slowed a little so hitting a fast-moving ball is more difficult than earlier. But even nonathletes can jog, swim, and take part in exercise. The current emphasis on fitness among adults may serve as a good example for their children who have been receiving insufficient exercise. Half of all young adults (70 percent of the college educated) are now exercising on a regular basis compared with only one quarter of the population in 1960 (Harris & Gurin, 1985).

Increase in height has practically ceased by the age of twenty but weight gains may increase even beyond the twenties. The maximum of strength and coordination is typically reached in the early twenties. Olympic athletes are predominantly in their late teens and twenties. Divers, swimmers, and sprinters are young adults. Typically, professional athletes—boxers, baseball, football, tennis, and basketball players—are beginning to be concerned about aging when they are past the mid-twenties.

In the first major study to examine the relationship between exercise and mental health, surveys of more than 32,000 adults from age 20 through old age were analyzed by the National Center For Health Statistics. It was determined that frequent exercisers (only about 15-20 percent of the adult population), who spend at least 30 minutes of vigorous activity three times per week had more positive moods and less anxiety than those who exercised less or not at all. Some of the possible reasons for these results are that exercise can distract people from their problems; offer them improved self-esteem; provide positive feedback about physical appearance; and release endorphins that generate analgesic properties (Stephen, 1985).

Nicholson (1980) suggests that becoming an adult entails a hazard not frequently considered. This is the fact that taking on a job, developing new friendships, and assuming family responsibilities is accompanied by giving up sports and games. But these young adults do not give up the caloric intake established in their more vigorous adolescent days. This causes many adults to become fat and it speeds their decline in general bodily functioning.

In the United States obesity is also a symptom of neglect and abuse. It is gaining recognition as a national problem. For two decades we have known for certain that obesity is associated with heart disease, respiratory problems, and lowered resistance to illness. The risk of diabetes is increased by seven times in the obese. Longevity is negatively correlated with obesity. Persons who are 20 percent overweight are at risk. The chances of premature death increase by 10 percent for every 10 pounds overweight (Turner & Aschner, 1985).

It is recognized that there are no simple explanations or single causes for the serious problem of obesity. The many probable causes are of different strengths in various individuals. Overeating during childhood is thought to generate more fat cells or more active ones. Others attribute obesity to heredity. Another theory is that obesity is a defense mechanism—persons overeat because they are under stress (or place themselves under stress by generating anxieties). The fact that there are some regions in which obesity is unknown, or at least is unusual, causes some to consider cultural factors. In Western, technologically advanced countries, overweight is common because of the sedentary nature of work. There is an abundance of fattening foods and a dearth of physically demanding jobs. Sedentary society, combined with widespread use of alcohol—which lowers inhibitions, including those which control food intake—makes obesity a readily sprung trap.

Anorexia and Bulimia

Anorexia nervosa (self-imposed starving) and bulimia (overeating followed by vomiting and purging) appear to be health hazards which are about as dangerous as obesity but not so widely publicized. Ninety percent of the victims are female between the ages of 15 and 25. It is estimated that about one percent of this age group is included. Ostensibly, anorexia stems from cultural pressure—"Thin is in." Most studies suggest that the cultural factor is less influential than personality disorder. The inability to face stress seems to be the dominant cause but the facade of fat provides an excuse (Garner & Garfinkel, 1985).

The half million victims who have the eating disorder called bulimia ingest large amounts of food (usually carbohydrates) in private and then purge themselves by using laxatives, enemas, diuretics, and self-induced vomiting. Bulimia comes from the Greek and means "huge appetite." Actually this is a misnomer because bulimics do not eat up to 50,000 calories a day because of hunger. Instead, they rely on food as a self-medication in the same way others use drugs to handle unmet personal needs. Most researchers believe that bulimia, like anorexia, masks personality problems of low self-esteem. But the personality traits associated with the disorders are different.

Anorexics shun food to cope with stress, withdraw socially, maintain rigid self-control, deny they have any problems and (despite a loss of up to 25 percent of their body weight) feel fat. In contrast, bulimics eat to cope with stress, are sociable, lose self-control, recognize something is wrong, yet because of the purging are able to maintain near normal weight.

The cure of these two diseases is not simply a matter of controlling one's eating habits. Self-image and confidence, anxieties, fears and goals must be basic considerations. The difficulty is shown by the fact that about 25 percent of anorexics will be helped, about 25 percent will not be helped and about 50 percent will gain some control but be vulnerable to recurrence. Seriousness is also indicated by the fact that about 10 percent of new cases will result in death (Swift, 1985).

The more successful approaches to these disorders are individual and group therapy in which victims can discuss causes, contributing factors, personal behaviors, and distorted perceptions. Because bulimics are usually extroverted and sociable, they respond best to group therapy, learning that they are not alone in suffering. Family therapy is usually required to assist anorexics. Local chapters of Overeaters Anonymous and Anorexia Unlimited may be the impetus for learning to control weight and achieve rational thinking about one's problems.

Diet and Disease

The U.S. Public Health Service estimates that 1 of 3 Americans born in 1985 will some day suffer from cancer. Adults can enjoy better health when they observe the following dietary guidelines recommended by the American Cancer Society.

1. *Avoid obesity.* Men and women who are 40 percent overweight increase their chances of colon, breast, and uterine cancer by nearly 50 percent. One hypothesis is that hormones produced by the fat cells may stimulate cancer development. If most people were to reduce their food intake by one third, they would be more healthy. It also makes sense to eat smaller amounts more frequently.

2. *Reduce fat intake.* The average intake of 40 percent of total calories should be cut back to 30 percent. In terms of choices this means more low-fat milk, fish, chicken, fruit, vegetables, and whole grain cereal. The eating of meat should be minimal and if possible eliminated.

3. *Increase high fiber foods.* There is disagreement about the influence of fiber. Some studies show a low incidence of colon cancer in populations whose diet consists mostly of unrefined food high in fiber. One view is that a lack of fiber slows down the transit time for body wastes. As a result, potential carcinogens in the waste are in contact longer with the membranes.

4. *Eat food rich with vitamins A and C.* Several investigations suggest that carotene, a form of vitamin A, can lower the risk of esophagus and larnyx cancer. Spinach, carrots, cantelope, peaches, apricots, and tomatoes all contain carotene. However, researchers caution against vitamin A supplements which can be toxic and lethal in large

doses. By inhibiting the formation of dangerous chemical compounds called nitrosamines, vitamin C is thought by some scientists to reduce the possibilities of cancer of the stomach and esophagus.

5. *Cruciferous vegetables belong in the diet.* There is a correlation between respiratory tract and gastrointestinal cancer and diets low in cruciferous vegetables like broccoli, cabbage, cauliflower, and brussels sprouts. Besides vitamin A, all of these vegetables include substances called indoles which have proven to be effective inhibitors of tumors in animals.

6. *If you drink, use alcohol in moderation.* Cumulative studies over two decades have shown that alcohol, when combined with smoking, increases the risk of cancer of the esophagus, larnyx, and oral cavity. It is believed that alcohol acts synergistically with inhaled cigarettes. Alcohol contributes to cirrhosis of the liver and tumor growth.

7. *Minimize salt-cured and nitrate-cured foods.* Because of incomplete combustion, smoked foods absorb some tars containing carcinogens similar to those in tobacco smoke. For this reason the practice of preparing food by barbecuing should be limited. The danger of consuming too much salt-cured food is suggested by population studies of high stomach cancer rates in countries like Iceland, Japan, and China where this type of diet is common (Kim, 1985; Burish, Levy & Mayerowitz, 1985).

Sexually Transmitted Diseases

During the 1980s the incidence of many sexually transmitted diseases (STDs) has declined (Holmes, Quinn, & Cory, 1985). One exception is AIDS (acquired immune deficiency syndrome) which is doubling every year. The U.S. Public Health Service predicts 270,000 AIDS cases by 1991. The only known ways to transmit the virus, which destroys white blood cells that protect the body from infection, include sexual intercourse, sharing the use of a hypodermic needle, and by transfusions of AIDS-contaminated blood. The virus has been found in tear and saliva.

About 75 percent of the AIDS victims are sexually active homosexual and bisexual men. The likelihood of infections is increased dramatically by exposure to multiple partners. That gay men have many more partners than heterosexual men has been known for some time. For example, most of the young adult gay respondents in a San Francisco study reported having 100 or more different partners in their lifetime. It has also been determined that multiple partners are common among older gay men. While any effective program to control SDTs depends on the identification of infected partners, this presents a unique problem in the gay community. It is not always possible for a patient to bring in his partner; he may have had contact with a number of strangers (Berger, 1983).

MIDDLE ADULTHOOD

Functional Physical Efficiency

Birth, rapid growth, youthful vigor, adult proficiency, and late-life decline is the characteristic pattern for all higher animals. With the passage of time normal human cells eventually lose the ability to function and divide. Human lifespan potential has not changed since ancient times. What has changed is the ability of more people to take advantage of scientific discoveries in regard to capitalizing on potential. The big, unsettled questions for humans in developed nations are how best to achieve and maintain peak proficiency and modulate the rate of decline. These is a growing consensus that humans are more likely to rust out through disuse than wear out via aging. This is apropos whether we speak of mental, physical, sexual, or social behavior.

Anthropologists have reported that the psychological implications of middle age are conditioned and changed by such things as mores, attitudes, economic conditions, and wars. In American culture, high value is placed on youth, physical beauty, and sexuality. There continues to be strong resistance, conscious and subconscious, to admitting one's chronological and functional age after about age 40. Fortunately, research indicates that aging is increasingly being regarded as a normal rather than alien process (Butler & Gleason, 1985).

The percent of total body weight that is muscle and bone declines during adulthood and the percent of the total that is fat increases. This means that the man who weighed 180 pounds when he was in college can at the age of 45 still weigh 180 pounds but he is fatter. This is an inevitable change but the rate of change can be slowed. By observing health rules—regular exercise, reduced intake of meat and sugar, and increased intake of fibrous foods—one can maintain the degree of wellness that slows that rate of bone and muscle loss. Not only does such precaution retard bone and muscle loss and reduce the incidence of obesity, it is claimed with research support, that the chances of coronary disease, diabetes, and cancer of the colon are reduced. The implications of physical development and wellness during adulthood are applicable to all age stages; namely there is a significant and partially controllable relationship between health maintenance and successful aging.

Sensory Acuity

Surprisingly close to age 45 (not so frequently needing the cautious phraseology of "about age 42 to 50") the phenomenon of presbyopia occurs. This means that fine print must be held at arm's length to be read—one is farsighted. The middle-ager asserts that "Phone directories and restaurant menus are printed in smaller type." The cornea tends to lose some of its transparency, admits less light, and demands better illumination for good vision than was the case in former years. The retina undergoes changes that reduce visual acuity, increase the problem of adjustment to darkness, and limit peripheral vision. There is a gradual decline in the eye's focusing ability. In short, middle age is usually accompanied by progressive but not necessarily handicapping deterioration of visual contact with the environment (Wilson, 1983).

There is also a gradual loss of hearing acuity, typically occurring first in the higher tones. Music becomes less and less enjoyable for some. It is vexing to listen to people with high-pitched voices because their enunciation seems to be "mushy." Acuity for low tones (and normal voices) declines less rapidly but there is some deterioration. This occurs much more frequently in men than in women. To some extent such a loss can be compensated for by sophisticated hearing aids; through an aid that merely increases volume may cause the wearer to be "blasted" by the also readily heard low tones. Cultural tolerance for hearing aids has not equalled that for glasses; but increasingly middle-agers are resigning themselves to their use—and to the admission of their age.

Sexual Function and Behavior

Inability to have children is about as clearly evident in middle-aged women as was their coming to maturity at the time of pubescence. The onset of adolescence was the first menstruation and the end of the childbearing period is marked by the less abrupt menopause.

On the average, menopause usually occurs between ages 45 and 50. The hormonal changes that cause or accompany the cessation of menses are often characterized by "hot flashes," sweatiness, weariness, irritability, faintness, and change in appetite. Normally women accommodate to these changes in stride. The unpleasant symptoms are ignored. Most are relieved that childbearing days are over. Seventy to eighty percent of women have no, or few, menopausal problems (Eisenberg & Eisenberg, 1985).

Men also undergo physiological changes that decrease the chances of further paternity. There is a continued decline in virility and a progressive decline in their active sperm count. The condition, known as climacteric, is noticeable when the male is unable to have an erection or to maintain it during intercourse. It should be noted, however, that anxiety, boredom, and misuse of alcohol also contribute to such a condition. The climacteric age varies, beginning in the late 40s for some men, with a small percentage not reaching it until about age 70. The menopause and climacteric can be modified or delayed through hormonal treatment. Such treatment, of course, must be given by a medical doctor who has some special knowledge of the interactive, pervasive nature of hormones. Hormonal treatment may affect digestion, activity pattern, beard growth, heart rate, blood pressure, kidney function, skin texture, and possibly cause cancer (Leiblum & Pervin, 1985).

Kelly (1980) states that the majority of sex problems are psychological: ignorance, boredom, anxiety, preoccupation with other events, and marital condition. By about age fifty most people are forced to admit their physical limitations. Organs atrophy, cell reproduction slows, and glandular secretions diminish. But this does not mean the cessation of sexual activity. The menopause does not have an adverse effect on sexual and orgasmic ability. Males and females in good physical and mental health can look forward to continuing sexual activity—with decreasing frequency—to and through old age.

There are indications that the middle-aged are handling menopause, climacteric, and empty-nest stage better today than in the past. These phenomena are not so threatening as they were 10 to 20 years ago. More people have gone through these experiences and have pioneered a variety of ways to make middle life enjoyable and productive.

The absence of dependent children provide middle-agers with income, space, and time they did not have with elementary, high school, and college-age dependents. Travel, smaller homes (often condominiums and apartments requiring less maintenance), pursuing a delayed college degree, doing volunteer work, developing hobbies, and getting to know each other again suggest that middle age is an opportunity rather than an inevitable deprivation. The potential benefits of an empty nest show the phenomenon of physical and psychological interdependence. This time of life can be stressful for those who are unprepared for it or who had stormy marriages. For others it is a time of tremendous personal growth.

Altogether it seems that sexual health has some similarity to learning and cognitive ability; namely, belief, attitude, self-concept, and behavior are more likely to be factors in the decline of healthy, normal, sexual performance than physical change alone. Evidence supports the need for thinking of physical, sexual, and mental change in terms of similar, but idiosyncratic rates of change.

LATER ADULTHOOD

Potential Lifespan

One view of aging is that the lifespan of an organism is programmed in cells. Geneticists postulate that aging varies according to the species under consideration. Each different animal reaches its prime, matures, and dies at rather precise times. Elephants, horses, dogs, and other animals have different lifespans. Just as an organism inherits eye color, it also inherits a genetic potential lifespan. Other factors (disease, accidents, psychological orientation) can influence the genetic program. There is evidence that the natural lifespan for humans is about 100 years (Gale, 1985). Although few persons will reach this distinctive age, most of us can expect to live longer than people in the past. Of course, there are subgroup variations which require continuing concern. White and black men respectively have shorter average lifespans (71 and 64) than their female counterparts (78 and 72). For both sexes on the average blacks die younger than whites (Schwartz, 1984).

In addition to genetic influence, the human potential lifespan is also related to physiology (Aiken, 1982). As people grow older they become less active because their body has less reserve to call upon. In effect, the organism gradually but inevitably wears out. The shortcoming of the theory is that cells do, to some extent, replace themselves. As we find antidotes to cellular wear, we may be able to lengthen the lifespan. It is possible to extend the lifespan by reducing strain and avoiding excessively hard work. Homeostatic imbalance seems to occur more easily in older than in younger adults. For example, the inability to maintain blood sugar or body temperatures. Homeostatic imbalance, like physiological deficit, may be aggravated by external factors.

Another postulation is that aging results in the buildup of metabolic wastes; e.g., calcification in arteries and increasingly brittle bones. The cross-linking theory assumes that connections between molecules contribute to lessened elasticity of tendons, skin, and blood vessels. Another theory postulates that the immune systems—the disease fighting functions—become weak or abnormal. The immune systems lose the ability to distinguish between self and foreign proteins. Arthritis, cancer, vascular diseases and hypertension have been associated with deficient immune reactions.

Physical Changes in Aging

Less is known about physical change with aging than has been thought because early studies so frequently consisted of institutionalized subjects. Now we know that such persons were atypical. Those in homes for elderly people and those on welfare constituted a rather special group. The stereotypes of aging, therefore, seem not to have accounted for those old persons who are active, alert, vigorous, and healthy. The stereotyped impression of later life is that people are weak, infirm, feeble, and afflicted with recent-memory defects. The view that most older persons are helpless, dependent, and emotionally useless is erroneous. Their loss of robustness is magnified in an achievement and action-oriented society. The truth is that less than 5 percent of persons over the age of 65 require some degree of custodial care. Even among those who are institutionalized, only about half of them have been diagnosed as being senile. There should be fewer cases of dependent seniors as the science of geriatric medicine becomes more widely available.

Less change occurs between ages 65 and 75 than between 15 and 25 or between 35 and 45. Nevertheless cumulative changes continue and differences already wide, become greater. Such changes that do occur derive from both physical and psychological factors. There are some age-specific changes but they may occur in different persons at widely spaced beginnings; e.g., wrinkling of the skin may be apparent in one person at age 45 but in another not until age 70. Visual adjustability decreases notably after about age 45. Visual defects continue to be cumulative, although again, some persons in their seventies have remarkably efficient vision. Hearing difficulties begin for some in middle age, and they rise in frequency as people become older. By the age of 70, about one in three persons has a significant hearing loss (Bennett, 1985).

Problems of diet become increasingly significant in later adulthood. There is a decline in resistance to stress with age. Older people take longer to regain their autonomic balance after psychological stress than do younger people. Exercise, neglected by many adults, is even more important in later life. Cardiovascular health is enhanced by regular, individually planned exercise. It also results in improved mental and emotional functioning for heart patients (Harris & Gurin, 1985).

One result of advancing medical technology and its availability in Western culture is that more people reach later adulthood with fewer after effects from prior illnesses. This means also that healthy persons in middle life reach later life as relatively healthy old persons. The elderly visit doctors more often than do young people and they spend more time per illness in the hospital.

The data relative to motivation and wellness are abundant in the field of geriatrics. More and more frequently it is reported that old persons who appeared to be invalids dying in nursing homes, or when living alone, have regained age-appropriate health after they again become involved with others. When a son, daughter, granddaughter or foster child have shown interest, love, and shared some time with the distressed person, that person's zest for living was revived (Jenkins, 1985).

Chapter 3

DEVELOPING INTELLECTUAL ABILITIES

After years of study about how the brain functions, our knowledge is still, at best, rudimentary. We are only beginning to appreciate the intellectual potential that can be developed through judicious and continuous learning. Although we have experienced the advantages and disadvantages of so-called intelligence tests throughout this century, a satisfactory and acceptable definition of intelligences (yes, there are more than one) has not yet been formulated. Nevertheless, enough is already known to provide some sound, even if tentative, guidelines for improving mental capacities.

In this chapter we consider the current "state of the union" regarding the brain and its intelligences. It is fair to use the word union because the brain is only part of a system. During the first century B.C., Lucretius, in his *On the Nature of Things*, wrote:

> The Mind which oft we call
> The intellect, wherein is sealed life's
> Counsel and regimen, is part no less
> Of man than hand and foot and eyes are parts
> Of one whole breathing creature.

True, the brain is only part of a human being but a truly marvelous part—and could be awe-inspiring if we make optimum use of it. In pursuit of this broad goal, let us examine (1) how the brain changes over time; (2) ways to assess its performance; and (3) methods for supporting individual achievement.

INTELLIGENCE AND THINKING

Changing Views of Intelligence

Historically, intelligence has been seen as a unitary and unifying concept of how humans make adjustments; e.g. among other definitions, intelligence is: the ability to make facile, appropriate, and unique adjustments to, and alterations of, varied aspects of the environment. It was apparent to pioneers of mental assessment that (1) the tests depended on an author's concept of intelligence; (2) tests only sampled (approximated rather than measured—as with a ruler or scale) intelligence; (3) human intelligence is constantly changing with age and circumstance; (4) at best, mental tests give tentative and suggestive results and need supplementary evidence (e.g., success in school or life) in order for them to be optimally helpful.

The concept of multiple intelligences has a rather solid basis in psychological history. Many early scholars and test designers rejected the idea of a unitary intelligence. Binet and Simon (1905), who are credited with developing the first mental tests, believed that intelligence was not a single mental ability but rather an aggregate of functions involving judgment, comprehension, direction, and invention. Thurstone (1938) postulated that there were seven primary mental abilities—visual, perceptual, numerical, verbal, reasoning, remembering, and problem solving—plus what he called a G (or general) factor that pervaded the others. He designed mental tests to probe these areas. Wechsler (1944) devised widely known and still used scales of mental abilities—one for children and one for adults. Both of these are available in verbal and nonverbal forms of cognitive response. The Stanford-Binet tests, devised by Terman (1937) have both verbal and nonverbal mental tasks. Perhaps the most inclusive concept was proposed by Guilford (1981) who suggested there were 120 discrete intelligences—most of which (but not all) can be assessed somewhat acceptably, by existing tests.

A recent attempt to modify the concept of intelligence has been made by Gardner (1983; 1985). He proposes that the major intelligences are linguistic, logical-mathematical, spatial, musical, bodily-kinesthetic, interpersonal, and intrapersonal. Pragmatically we recognize the varied aptitudes persons have for interpersonal relations. Gardner is the first to call them intelligences. The intelligences he cites involve the cultivation and use of information processing, problem solving, and development of talents. Students and those who guide them can be involved in deciding whether to use talents or let them atrophy, to develop a level of utilitarian competence, or cultivate notable skills in the various areas.

Another view regarding varied intelligences is that contributed by Sternberg (1986). He proposes that conventional mental tests do a disservice because they are limited to a particular aspect of total cognitive potential; i.e., componential intelligence, which is the kind of quick recall of facts, words, and ideas of others. Another kind of intelligence is what he calls experiential. The possesser of this kind of intelligence may or may not have high conventional test mentality; but combines disparate experiences into a meaningful relationship. This is often called insightful or creative response. The third component of intelligence (again, not assessed in typical mental tests) is call contextual. The possessor of this intelligence he calls streetsmart. These people know how to play the game of life and use their perceptions to manipulate people and things. Sternberg's triarchic theory fits admirably in the discussion, later in this chapter, on learning styles and special abilities. An attractive feature of Sternberg's postulations is his belief that these aspects of intelligences can, by appropriate teaching and experience, be strengthened and improved.

Implications of Varied Intelligences

Taylor (1968) suggested two decades ago that there could be a "New Dawn" in education if schools would use many tests to assess multiple intelligences. Then instead of having groups of children labeled as bluebirds, robins, and sparrows—assembled according to mental test scores, each student's unique strengths would be identified and given initial or primary attention. The relative weaknesses would be considered after students had a chance to demonstrate ability and experience success in their area of strength. In practice educators typically do the opposite—they take the remedial approach, try to bring the student "up to average," and thus label the individual as deficient. By starting with the area of strength a student can experience success, be more likely to enjoy school, and establish a base for interest in lifelong learning.

Brain Hemispheres and Learning

There is an exploding amount of information about brain dynamics that cannot be fitted into the age-stage manner of presentation. Part of these new data refer to the fact that the human brain is not symmetrically balanced in terms of function. This is true from birth to old age. Physically, the right half of the brain is a sort of reflection of the left half; but what each part does is distinct, although complementary. It seems that marked improvement in the way the brain is used—in schooling, work, and life—could be achieved if teachers and other adults were to recognize the special functions of the right *and* left halves of the brain (Bradshaw & Nettleton, 1983).

The brain is covered with a wrinkled layer called the cerebral cortex. This cover is divided by a fissure running from front-center to rear center. The halves are called hemispheres. Neurologically, each brain hemisphere controls the opposite side of the body. The importance of the "divided brain," however, goes far beyond this crossover phenomenon.

It has been suggested that each of the two brain halves might be capable of working independently. For example, in one experiment a composite photograph was made, consisting of half the face of a young boy and half the face of an old man. Subjects were seated so they were at the midpoint of the visual field. The photo was flashed for one-tenth of a second—too fast for the eyes to scan and adjust cooperatively. The impression received through the left eye was carried to the right half of the brain, and the impression received through the right eye was carried to the left. When subjects were asked to tell what they had seen, they described the old man. When asked to point to what they had seen, however, they pointed to the full photo of the young boy. The subjects were unaware of the inconsistency in their responses (Restak, 1985).

Sperry (1982) won the Nobel Prize in Medicine for his work on the split brain. His efforts and that of other researchers implicate the left hemisphere as being most involved with language functions, abstract and analytic thinking. It is particularly active when a person engages in logical, verbal, or mathematical functions. This could be called the scientific mind, the part of our brain that is given almost exclusive recognition in school work. In contrast, the right hemisphere seems to control spatial perception, musical and artistic skills. It is more active when one engages in visual perception, intuition, and artistic production. Given these special capacities, it is easy to see why the left hemisphere has been referred to as the "rational" half of the brain and the right hemisphere as the "intuitive" half. Yet there is no evidence that people are purely left brained or right brained. Neither is it true that one hemisphere can be educated at a time while the other merely idles along unconsciously. Instead, we have one differentiated brain which integrates the contributions of both hemispheres to enhance human thought and well-being.

Up to this time few teaching implications have derived from knowledge about the split brain. But we know enough to begin some deviation from conventional practices. For example, we can acknowledge that intuitive and creative abilities of students merit attention in the classroom. These can be the starting point for developing interest in school subjects. So much of education is based on the left, and logical hemisphere and so little is based on the right, or intuitive and creative hemisphere that some students have "no place to go" (Wittrock, 1985).

Age-Stages in Cognitive Development

Throughout this book the continuity of development has been emphasized; i.e., development today depends on what occurred previously and what happens today influences or shapes the future. However, this continuity is not a simple, steady flow. There are some time frames—as our chapter subtopics indicate—that contribute more to a given phase of development than do others. The phenomena of readiness (e.g. reading readiness), critical periods, atrophy of potentials through nonuse (e.g. loss of creative potential by many elementary school students) show that age-stages as well as continuity are significant developmental factors.

Jean Piaget (1926; 1954; 1970; 1981), a Swiss epistemologist, is the best known proponent of age-stages in cognitive development. His early publications received little attention in the United States but during the 1950s his ideas literally revolutionized our thinking. He strongly influenced psychological theory, experimental cognitive studies and was eagerly accepted in modifying classroom practice. Since 1970, books and journals on teaching methods have contained numerous descriptions confirming, questioning, and expanding Piaget's evolving theory of age-stages in cognitive development.

In Piaget's view the best way to help students learn is by presenting them with curriculum they can understand at their level of thinking. It is erroneous to suppose that a child's thinking is essentially like that of an adult, any differences being due simply to the child's lack of experience. Children are not miniature adults, and at each stage their thought is both distinctive and different from that of grownups. Piaget asserted that children's thinking evolves through four specific stages, roughly associated with age, but not totally independent of individual differences, experience and affective factors. In turn, each stage will be presented in relation to the age group it describes: Sensorimotor—Infancy; Preoperational—Early Childhood; Concrete Operational—Middle and Later Childhood; and Formal Operations—Adolescence.

INFANCY

Nutrition and Brain Development

The time during which growth of an organ, or organism, is most rapid may be considered to be a critical period. Two critical periods for brain development occur between conception and a child's second birthday. The first period of five weeks is between fetal age of eight and fourteen weeks. At this time of spurting brain growth, billions of cells called neuroblasts develop into neurons. Each individual has a full complement of neurons by 13 or 14 weeks. Nourishment during this time influences the total number of neurons one possesses. It is estimated that severe undernourishment, common in developing nations, may result in 40 percent fewer neurons than in well-nourished Western babies.

The second critical period of brain development begins about 10 weeks before birth and continues to about the baby's second birthday. The focus of this critical issue is not increase in number of neuroblasts but the environment and experiences that initiate and exercise the complicated interconnections of the neurons (Caplan, 1982; White, 1985).

Cumulative evidence points to the hypothesis that low socioeconomic status in Western culture contributes to failure to capitalize on brain potential. Poverty exerts its toll from the moment of conception. For instance, there are more stillbirths and a greater incidence of prematurity among the poor than among those of higher socioeconomic status. There is ten times the chance of mental retardation with prematurity than in the full term infant (Schwartz, 1984). Several factors converge to make lower socioeconomic women "reproductive risks," i.e., comparatively inadequate mothers.

1. They themselves are often undernourished or improperly nourished—eating insufficient protein and being vitamin deficient. This finding obtains for cities and rural slums in North America and Europe.
2. Whether married or unmarried, lower socioeconomic status mothers are younger than those of higher socioeconomic status families.
3. Lower socioeconomic status mothers have less formal education than those of higher socioeconomic status. They less often take advantage of prenatal instruction and prenatal clinics—if such are available.
4. Syphilis, contracted pelvis, and malnutrition contribute to premature labor and difficult births. All these difficulties are inversely related to socioeconomic status.

Comparative studies on laboratory animals fail to give much hope that postnatal care can fully compensate for prenatal deprivation or damage. Malnutrition of pregnant women results in smaller fetal brain cells, but the fetus draws from the mother and the deficiency may be reversible. After birth, during the first six months, malnutrition retards cell division and the consequent lessened number of brain cells results in permanent impairment of learning ability (Kim, 1985).

Malnutrition is now recognized as a major factor in child morbidity, mortality, and failure to achieve normal growth. To maximize brain growth through the critical period from conception through about six months of infancy, the following considerations are focal. Prospective mothers should avail themselves of existing dietary and medical clinics, seminars and counseling. This should help keep them in good health as they give attention to an adequate exercise and diet in accord with medical advice. Mothers should avoid the use of tobacco, alcohol, and drugs other than those medically prescribed. After birth infants must receive the nutrition, including vitamins and proteins, and emotional support that provide optimum development of the nervous system.

Sensorimotor Thinking

According to Piaget (1970; 1981) the first stage of thought is the prelanguage or sensorimotor stage. In this stage, between birth and age two, children understand the nature of things only in the context of their sensory, physical presence. What they learn about objects is based largely on perception and motor activity; looking, feeling, tasting, pushing, banging. The emergence of intelligent behavior is revealed by eight or nine-month-old children when they search for an object that has been removed from immediate sensory perception. Prior to this development, when an object is removed children act as though it no longer existed. Then objects come to have permanence, whether or not they are in the immediate perceptual field. Piaget says they are invariant. When children realize that things are real regardless of location the next stage of cognitive development can be identified.

EARLY CHILDHOOD

Brain Growth Spurts

Epstein (1978) was the first researcher who called attention to the fact that children's brains grow by spurts:

3 months to 10 months
2 years to 4 years
6 years to 8 years
10 years to 12 or 13 years
14 years to 16 or 17 years

The reason why the phenomenon of brain growth by intervals had not been reported earlier is twofold: (1) Most researchers on children's brain growth grouped together too wide an age range. (2) The researchers ignored the brain weight differences, for example between 2–4 years and 6–8 years, because they were quantitatively small. But says Epstein, these small differences are highly significant. These "... brain growth stages are not a theoretical notion but a scientific fact ..." (Epstein, 1978, p. 345). He presents studies to support the assertion.

Among the supporting data is an argument that generated much contention in the psychological community during the 1970s. Jensen (1969; 1980; 1984) presented data which upset prevailing beliefs about the merit of compensatory education. He denied the value of trying to compensate for cultural differences through kindergarten and preschool education for children of low socioeconomic status, especially nonwhites. He claimed that because intelligence is about 80 percent inherited, compensatory education would be ineffective. His data seem to prove the claim. Most often the intervention effects did not last.

Some scholars contended that Jensen's data were either incorrect, specially selected to prove his point, or that his inferences were wrong. Still others argued that if only 20 percent of one's intelligence were environmentally influenced even that much was worth working on in compensatory education.

A decade later Epstein (1978) cast new light on the fascinating arguments. He states that Jensen, his supporters, and the many who argued against him would have been less acrimonious if they had debated from a different base; i.e., spurts of brain growth. Epstein contends that Jensen made an error in supposing that the only interpretation of his data on the ineffectiveness of preschool education was a genetic difference. The error made by Jensen and those who disparaged his conclusions was that preschool programs had, in the majority of cases, focussed on the wrong age. Look again at the brain growth by spurts data. Preschool programs had concentrated on exactly the wrong ages for intervention (Epstein, 1978). Ages four to six are fallow years and because the brain is most responsive during the critical periods of rapid growth, remedial programs should have selected ages two to four or six to eight. This is what Jensen and those who did conduct successful programs found; but they did not recognize the significance of small differences in ages. Work with two- and three-year-olds was successful; but when the data were thrown in with data for four- to six-year-olds the advantages for two- to four-year-olds became obscure. (The error that occurs from including too broad a spectrum will be noted again in the section on mental testing; i.e., include too many subaspects of intelligence, and the prominence of one subdivision or talent is obliterated.)

Epstein's conclusions are similar to those reached by Restak (1985). Brian cell division is complete before two years of age; but brain weight continues to increase as cells become larger and thicker by adding protein, ribonucleic acid (RNA), and lipids. At this point the issue is equivocal because cause and effect become inseparable. Researchers do not know whether experience triggers protein growth, RNA gains, and lipid changes, or whether experiences become more meaningful because the cells have grown. But the implications are clear. Intentionally enriched environments and experiences for two- to four- year-olds do produce marked and enduring gains in intelligence. Indeed, Pifer (1983) cites longitudinal evidence that early intervention still pays off twenty years later. The 123 experimental toddlers from low-income families in Michigan who attended a program during the 1960s were, in contrast to the controls (nonattenders), two decades later more likely to finish high school, become employed, go to college, avoid teenage pregnancy or conflict with the law, and score higher on achievement tests.

Preoperational Thinking

This stage, which Piaget (1970; 1981) says reflects representational thought, begins at about two years of age and ends at about six. Although still guided by appearance of things, preschoolers can now represent objects symbolically through imitation, drawings, playthings, and language. They learn to sort objects in terms of a single factor such as color, size, and function. At this stage, children's understanding includes the invariant existence of an object—its permanence—but not conservation. The term *conservation* here refers to an understanding that, unless something is added or taken away from an object, its quantitative aspects remain the same despite any external change in appearance.

Because preschoolers rely more on perception than reason, they are easily misled when confronted by one or another of the Piagetian conservation tasks. Consider the conservation of substance. Before age six or seven a child will observe that two balls of clay are of equal size and then declare that the piece reformed into a long, thin strip has more clay than the ball-shaped piece. Similarly, preschoolers believe that the same amount of liquid is greater when it appears in a tall, narrow container than when it appears in a short, wide one.

This failure to understand the conservation of volume attributes in part to the characteristic known as *centration*—focusing attention on just one aspect of a situation at a time. The preschooler tends to focus on either height or width of the container and fails to note that the other is also changing in a compensating way so that volume remains the same. Another limiting characteristic involves being unaware of *reversibility*. It is clear to older children that the liquid could be poured back into the original container and would then take up the same space as before; its volume has not changed. But to the preoperational child, the change was more basic; he is unaware of reversibility. Consequently, these features of thinking combine to prevent the recognition of conservation. For a less formal assessment of this cognitive limitation, ask a four-year-old if he has a brother—"Yes." Then ask: Does your brother have a brother?—"No." This is irreversibility.

The thinking of preschoolers is also limited by egocentrism. This is not a derogatory term but instead describes an excessive reliance on the child's own point of view with the consequent inability to be objective. He finds it difficult to comprehend how anyone can see things from a different viewpoint and as a result he is "always right." This presents predictable problems during the inevitable conflicts which accompany dominion play. Piaget observed that nearly half of the preoperational child's speech is egocentric. For example, collective monologues can be observed in preschools whenever several boys and girls are engaged in common activity. All of them are talking about what they are doing but none listen to others; all are simply talking aloud in front of one another.

MIDDLE AND LATER CHILDHOOD

Concrete Operational Thinking

During the period from about age six to eleven children gradually attain the ability to conserve each of the quantitative aspects of objects—mass, length, number, area, weight, and volume. Piaget (1970; 1981) suggests that reversibility of thought is the most important mental operation differentiating elementary students from preschoolers. Reversibility permits an individual to carry thought backward as well as forward, a requirement for arithmetic reasoning because addition and subtraction are the same operation carried out in opposite directions. When faced with the liquid conservation task mentioned earlier, children at the concrete operational stage use reversibility; that is, they pour the liquids back into their original containers to determine whether the two quantities take up the same amount of space.

The concepts of conservation and reversibility take on additional meaning as children mature. In the concrete operations stage children begin to apply logical thought to specific problems. When faced with a discrepancy between thought and perception, as in conservation problems, they make logical rather than perceptual decisions. As logic is acquired, boys and girls move from egocentric thinking toward objectivity. They no longer see the world exclusively from their own vantage point; they are capable of looking at situations from another person's point of view.

Styles of Learning

By the time children enter first grade, their brain is 90 percent of its full size. At school attention is devoted to developing the greatest of all talents, learning how to learn. This goal is not restricted to elementary school but during middle and later childhood many of the basic skills and attitudes toward self and learning are being formulated.

Since colonial days, when students lined up in rows and recited lessons in unison, there has been a continuing aspiration to respect individual differences. Initially, this meant making some kind of provision for slow learners—special classes, varied assignments, retention in grade, and relaxation of requirements. For those who were rapid learners special promotions were available but relatively few special classes.

There is, however, much more to individual differences than just rate of learning. Much interest in differences focuses on those manifested in special talents and variations in learning styles (Dunn, 1984). If teaching at home and school is to be most effective parents and teachers will profit from recognizing and adapting to learning styles such as the following:

> Learners use varied sense modalities, i.e., some depend primarily on sight, some depend more on listening, and others may rely on doing, touching, and manipulating (feeling).
>
> Some learn most effectively from other children in groups; and differ further by being leaders, participators, or observers within the group (a manifestation of social intelligence). Some learn best from adults and others are most efficient when they are alone as experimenters, thinkers (reflective style), readers, or as observers.

Children differ in thinking format, i.e., some accumulate much knowledge and then organize it into patterns; others first get an overall picture and then study details.

Learning styles parallel lifestyles. Some persons are formalistic—their learning stems from structure, rules, authority, and established values. Others are sociocentric—they take their clues from peers, parents, and teachers. Still others have a personalistic style—they prefer independence, freedom, and self-direction.

Some learners are deliberate and reflective; others are impulsive, energetic, and intuitive.

Field-dependent learners rely on the situation and their colleagues and teachers to choose activities and answers. Field-independent learners prefer to make their own choices and draw their own conclusions.

Many children are comfortable with structure, regulated discipline, and assignments, field trips, and experiences determined by others. Such regimes are anathema to those who prefer permissiveness, risktaking, and self-determination.

Talking is a learning style as well as a teaching style. It is a style preferred by a much larger number of children than get a good chance to use it. [There is too much talking on the part of adults and too little dialogue with and among children.]

Bruner (1983) provides a theoretical reason for paying careful attention to learning styles as dictated by brain structure. Studies of brain localization provide a physiological reason for acknowledging varied learning styles because electrodes show different areas of development for language, motor, and sensory areas in different children. For example, talented children learn to play musical instruments very well and with surprising ease. Multiple research investigations should be recognized and, as far as possible, be used as a basis for teaching-learning transactions. If such recognitions of learning styles were implemented in homes and schools, there could be a revolution in children's development (Dunn, 1984).

Brain scientists are focussing an increased amount of research to pinpoint the groups of cells that are operative in specific cognitive functions. As knowledge accumulates we may, through X-ray scans of the prefrontal lobes of the brain, greatly increase knowledge of the neural bases of learning—memory, emotion, and problem solving. The use of drugs may become useful in the remediation of deficiency (Wittrock, 1985).

Recognition of all learning styles is difficult when dealing with classrooms filled with individual differences. Fortunately it is unnecessary to recognize every variant style in every student. Most children are pliable enough that they can, and have, through centuries adapted to whatever learning milieus were available. However, if learning styles were made the point of departure in innovative teaching of persons having difficulties we might do still better. We might raise that alleged 10 percent of the brain power persons have that is vigorously used to a somewhat higher percentage. Bruner (1983) and thousands of teachers believe that such improvement is a distinct possibility. Successful adults who were once potential dropouts, behavior problems, or suffered from boredom have demonstrated that even token recognition of their differences and learning styles can work wonders in motivation—and brain usage.

Creativity and Intelligence

In the early 1980s some researchers warned that America's prominence in technology could be sustained only if greater emphasis were assigned to creative thinking in elementary and high schools. It was further suggested that the ability to discover new relationships and generate new perspectives would require educators to give more attention to classroom activities favoring the right brain. Currently, however, the right and left brain dichotomy is regarded as too simplified and misleading in terms of promoting creative behavior. Instead, there is now recognition that creativity implicates both conscious and unconscious aspects of brain functioning. Because it draws upon intuition, logic, knowledge, and imagination (both sides of the brain), creativity should be seen as the ability to use different modes of thought. Indeed, Gardner (1985), who proposes that humans have at least seven distinct forms of intelligence, indicates that no single intelligence is intrinsically creative. Creativity requires honing one or more of our intelligences to a high degree.

Adolescence

Formal Operational Thinking

By the time of adolescence the brain has reached maximum weight. Although the brain is no longer growing physically, other changes are taking place. Piaget (1970; 1981) has shown that the nature of thinking shifts from the child stage called concrete operations (needing material objects to understand combinations, identities, and reciprocal relations) to an ability to think logically and abstractly.

This final stage of cognitive development called formal operations begins at about age 11 to 15 for some, in later adolescence for others, and for a certain segment never does occur. Formal operational thinkers need visual props less as they become able to handle abstractions. They become capable of forming hypotheses and deducing possible consequences. During this stage they also learn to use symbols for other symbols, as in algebra, and to handle proportions. They can apply reversibility in new ways—for example, comprehending that increasing weight decreases the distance needed to balance objects on a fulcrum, while increasing the distance decreases the need for weight.

The ability to conceptualize all the possible combinations in a system permits teenagers to consider contrary-to-fact propositions and treat them as if they were true. "Let's suppose snow is black" is a proposition some adolescents can accept whereas concrete operational peers would insist "Snow isn't black, it's white." The ability to take opposing viewpoints into account increases reliance on the scientific method, decreases self-centered thinking, and can improve interpersonal relations. Before high school, students consistently rank history near the bottom in curriculum preference. Only at the stage of formal operations do students begin to comprehend historical time. The distant past assumes meaning, possible futures are contemplated, and life in other cultures can be understood.

Teaching Implications of Piaget's Theory

Now that we have examined all of Piaget's stages, let us consider some implications and the concerns of scholars who question his conclusions. Piaget emphasized that children must perceive, talk about, and manipulate objects to develop intellectual abilities. First hand experience, however time-consuming, is the key to stable and enduring learning. What matters more than verbalizing rules or committing facts to memory is engagement in practical activities that call for problem solving. Consequently, a fundamental obligation of caretakers of any age group is to provide children with tasks that permit them to acquire understanding. One reason for Piaget's enormous popularity is that his conservation and formal operations tasks enable teachers to determine more rationally the kinds of assignments from which individual students can benefit. Some of his tests, however, are worded in such a way that children can become confused. In turn their abilities may be underestimated (Donaldson, 1978).

Piaget's theory emphasizes the futility of hurry and the wisdom of utilizing children's heterostatic thrust. Problems that are slightly puzzling, not devastating, are intriguing. The theory thus presents teachers with the continual challenge of individualizing learning programs. Although it points out the invariability of the developmental sequence, it recognizes that learning can be enjoyable when each child's level of development within an age stage is used as a guide in teaching.

Teachers would be wise to recognize Piaget's repeated emphasis that the line between stages is often blurred and partial. On the other hand, Niemark (1975) reports that only 40 to 60 percent of adolescents and adults perform at the level required for Piaget's stage of formal operations. One interpretation is that these findings have dire implications for high school systems which mandate that every student take three or four years of mathematics and science. It can be predicted that this unreasonable expectation will influence many teenagers to drop out of school. An alternative explanation is that Piaget's tasks pose too rigorous a standard. Common sense suggests that most people regularly engage in formal operations. We make plans knowing that unforeseen events may necessitate changing them. Then too, most of us routinely evaluate the logic of what we are told by others. Granting the limitations of his assessment methods, Piaget's leadership in the study of thought processes is without dispute.

Emergence of Cognitive Subabilities

The changing brain of adolescents was not the discovery of contemporary leaders in the field of psychological theory. Segal (1948) reported that several research studies pointed to the conclusion that cognitive development in adolescence changes from its earlier nature. Prior to the teenage years mental progress is of a general nature. As a global entity, intelligence has its maximum growth during childhood. But in adolescence specific subabilities of intelligence—e.g., spatial, verbal, memory, induction, deduction, computation—emerge with sharper clarity. Correlations between specific subabilities become smaller with the passage of years. Stated another way, the subabilities which comprise total intelligence become more and more distinct—independent from each other.

This datum has significance for the assessment of potential for classroom performance. All adolescents are not endowed with the same potential for competence in specific areas—verbal, mathematical, form perception—even when they have the same IQ. This fact is taken into account by tests that probe different aspects of intelligence; for example, the *Differential Aptitude Test*. If students and their teachers knew that one individual might perform with mediocrity in language classes but still might perform well in mathematics, there might be less discouragement regarding a low mark in one subject. There might be more inclination to exercise brain potentials instead of trying to bring someone "up to average" in the weak cognitive areas by placing the student in a remedial class. Such emphasis on weakness can and often does undermine student self-concept. This erosion of self concept might be a greater hazard than we realize if it is true that the major developmental task of adolescents is the search for identity.

Brain Reorganization

There is evidence that the brain undergoes more reorganization during adolescence than has previously been supposed. At UCLA neurologists use positron emission tomography (pet) scanners to measure biological activity of the brain and other organs. When a radioactive form of sugar injected into the bloodstream goes to the brain the more active cells metabolize greater amounts of sugar. During this process the pet scanner senses active versus inactive brain areas and provides a picture of metabolic activity. Infants show a metabolic rate about two-thirds that of an adult. By age two the rate is similar to grownups with rapid increases in the area where higher mental functions occur, the cerebral cortex. The brains of three- to eleven-year-olds use twice as much energy as the brains of adults. But then the metabolic rate gradually falls until it reaches the adult level at about age fourteen (Wittrock, 1985).

In a related study at the University of Chicago, neurologists found twice as many synaptic connections, places where the branches of brain cells meet, in the cortex of children than adults. The extra connections in a child's brain would utilize more energy as reflected by higher metabolic values. These data suggest that during early adolescence there may be an elimination of little used synapses resulting in a functional stabilization of the brain and lower energy demands. Consider one additional factor. Sleep patterns have been measured for all age groups at the State University of New York. Children from two to ten years of age get twice as much deep sleep as adults. But deep sleep declines by 50 percent between age eleven and fourteen. It seems reasonable that the more metabolically active child would require more deep sleep. And, if the adolescent brain has fewer synaptic connections, it would need less sleep. The fact is when children enter adolescence their total duration of sleep decreases by about two hours per night (Restak, 1985).

The relationship between metabolic activity, number of synapses, and relative amounts of deep sleep remain a matter of speculation. But the brain seems to develop all potentially useful neural connections by age two. During the first decade of life experience may strengthen the neural circuits that are used and ultimately eliminate those that are not (aided perhaps by changing hormones). When the ability to handle abstract problems emerges in adolescence, these skills are often seen as a product of continued learning. But maybe they develop only after the brain rids itself of excessive neural connections. Looked at another way, perhaps the inability of children to concentrate for long periods may attribute to having excessive neural connections that interfere with sustained logic.

Assesment of Intelligence

Lewis Terman (1937; 1960) of Stanford University devised a widely known and historically pivotal individual test of intelligence. He called it the Stanford-Binet intelligence test, in order to give credit to Alfred Binet of France, who had provided the theoretical base for the test. The scores were expressed in terms of IQ, which was computed by dividing a student's mental age (obtained from the tests) by chronological age—multiplied by 100 to avoid using decimal fractions. The formula expresses the idea that theoretically children grow at the same rate mentally as they grow older in years. This is somewhat true, for about the first 10 to 12 years, for a considerable portion of the population. Children who are above average in tested intelligence have a mental age that is higher than their chronological age (expressed in months) and hence have an IQ above 100. For example, a 12-year-old who scores as high on the Stanford-Binet as average 14-year-olds has an IQ of 116:

$$\frac{14 \text{ years (mental age)} \times 12 \text{ (months per year)}}{12 \text{ years (chronological age)} \times 12 \text{ (months per year)}} = \frac{168}{144} \times 100 = 116$$

Mentally slow children keep falling behind in mental age in relation to months they have lived so their IQ is below 100. For example, the 12-year-old who scores the same as typical 10-year-olds has an IQ of 83:

$$\frac{10 \text{ years (mental age)} \times 12 \text{ (months per year)}}{12 \text{ years (chronological age)} \times 12 \text{ (months per year)}} = \frac{120}{144} \times 100 = 83$$

Terman recognized that intelligence decelerates in growth rate earlier than do the number of birthdays we celebrate. In fact, we still do not know the age at which intelligence stops growing. To keep dividing mental age by chronological age for persons over 20 would make it seem that our entire adult population was feeble-minded. Therefore, in keeping with the postulation of his time, the highest chronological age Terman used was 16 (Terman & Merrill, 1937). It was, in other words, supposed that the capacity for learning, or intelligence, had achieved its full adult stature by about age 16 and had begun to slow its pace by about age 13. So, 14-year-olds did not have their mental ages divided by the full number of months they had lived. The IQs after age 13 were provided by a table of suitably adjusted computations.

Cessation of Cognitive Development

The brain has achieved its maximum weight by the time of late adolescence. Curves representing the growth of intelligences have depicted the belief in vogue during the first half of this century. That postulation was the one used in the first three editions (1914, 1937, 1960) of the Stanford-Binet tests, to the effect that cognitive growth reached its peak in late adolescence. Some believed that global intelligences (those evaluated by conventional tests) peaked at about age

13 (for those of lower IQ) to about 17 (for those of higher IQ) (Bayley, 1949; E. Thorndike, 1927; Thurstone, 1955). After cognitive development had supposedly reached its mature level some time in the teens, there was a plateau for peak level performance that lasted for a decade. Then a slow decline was alleged to occur.

Robert Thorndike (1948), the son of E. Thorndike (cited above) was one of many scholars who questioned the hypothesis regarding peak mental growth in the teens. He tested 1000 students from ages 13 to 20, using the American Council on Education Psychological Examination and reported an average gain of 35.5 standard score points. GIs attending college after military service made distinctly better scholastic records when compared to younger college students. The younger students had comparable mental test scores and school records but entered college immediately after high school. Of course, the advantage of the GIs might have been emotional maturation or motivation more than intelligences; but still the notion that after age 22 or so there is a gradual decline was questioned.

Development, defined as the improvement of function, may be contrasted to growth—the addition of mass, weight, or length. The brain may stop growing in youth but scholars believe that the age at which intelligence ceases to develop has in the past been placed much too early. The current prevailing view is that cognitive development, or decline, is more a function of lifestyle and personality than it is a correlate of age. There seems to be the distinct possibility that the timespan for the growth of intelligences is positively correlated with the vigorous pursuit of learning; e.g., with the continuation of purposeful study.

Identifying the Gifted and Talented

Programs for the gifted and talented have shifted in recent years to include a broader range of abilities. Although we do not know just how much of a person's potential goes undetected by intelligence tests, the available evidence suggests it must be considerable. This was apparent in a study by Guilford (1977) during which scientists were presented with a list of 28 mental functions and requested to rank order them in terms of relevance for successful work in the physical sciences. All but one of the traditional intelligence test factors ranked below twentieth on the list; that is, 19 out of the 20 characteristics considered by scientists most important on the job were not included on the traditional mental tests. The need for broader testing is reinforced by Lyon (1981), former Director of the Federal Office of Gifted and Talented. He indicates that the effect of considering only those persons with an IQ of 130 or above (the traditional definition using general intellectual ability) as gifted would result in overlooking 70 percent of the creative and talented. Consequently, the definition of gifted and talented has enlarged to subsume five categories: (1) general intellectual ability; (2) specific academic aptitude; (3) creative and productive thinking; (4) visual and performing arts; and (5) leadership ability.

The search for students who qualify in each of the categories is causing educators to look beyond the traditional procedures. For example, a two-step process facilitates the identification and placement of junior and senior high youngsters who are precocious in various academic areas. First, a rough guide to who is gifted derives from results on an age-appropriate achievement test for some particular talent, such as mathematical or verbal reasoning ability. Students who score in the 96th percentile or better are then given tests designed for older persons such as the College Board's Scholastic Aptitude Test (SAT). This second screening spreads out the tight cluster at the ceiling of age-appropriate tests and enables guidance in a proper enrichment or accelerated curriculum.

The use of learning style inventories is another means for detecting previously unrecognized strengths. Some of these inventories reflect research on the split brain. Because differences in student potentials are influenced by whether the right or left hemisphere of the brain dominates, Torrance's (1986; 1986a) tests deserve consideration. His instruments confront subjects with: (a) problems of a sequential logical linear nature that depend on left-hemisphere functions; and (b) problems calling for adaptation, intuition, and innovation which are right-hemisphere functions. Some problems in the test call for integration of the two sets of functions.

More sources of perception are being considered in determining who is gifted. An emerging practice is to augment the value of tests by inviting subjective nominations as well. This shift is especially relevant for the performing arts, creativity, and leadership categories. It has been obvious for a long time that expert judges in the performing arts can detect talent which is overlooked by the school. Similarly, youngsters who show leadership (even in negative ways like a gang leader or con artist) are more likely to be noticed by a source outside the school—parents. Because an often stated purpose of programs for the gifted is to promote leaderships, it would be well for educators to study, through reading and peer discussion, the identification of potential leadership.

Parent nominations closely correlate with success in gifted programs. Hopefully, the low reliability of teacher nominations will improve as colleges better prepare educators for the classroom. When children are asked to identify their outstanding peers, the observations are quite accurate. Even more reliable are the gifted themselves, who can tell which peers share their characteristics and belong in the program. There is also the occasional self-nomination, a person not recruited by others but capable and motivated. In combination these sources can help schools identify more students who ought to be in gifted programs (Harrington & Harrington, 1985).

When Congress passed the Gifted and Talented Children's Act in 1978, the federal government committed itself to improving instruction for this overlooked group. Unfortunately, by 1983 funds were no longer available so most efforts at curriculum reform and programmatic innovation had to be discontinued throughout the country. The tragedy is underscored by the fact that nearly 20 percent of all dropouts in the United States are gifted, a far larger proportion than their representation in the general population.

In a comprehensive survey of 1,200 school districts Cox, Daniel, and Boston (1985) found that more than 70 percent of schools offer only part-time "pull-out" programs for gifted students. By this arrangement boys and girls leave the regular classroom for a limited period of time each day or once a week to study with peers of their ability level. Some of the recommendations by the research team include:

> Abandon pull-out programs in favor of full time curriculum offerings for able learners.
>
> Establish a flexible flow of curriculum so that students can progress at their own pace.
>
> Continue to broaden the criteria for selecting participants in able learner programs.
>
> Eliminate the term "gifted" as a label because students are frequently able in a particular arena of learning but not necessarily in others.

Early and Middle Adulthood

Creativity and Job Success

One point of view about the age-creativity relationship implies that there is a biological "time of plenty," an age period during which inventive behavior will most likely flourish (Baldwin, Colangelo, & Pittman, 1984). Advocates of this position insist that we can infer distinctions in the development of creative thinking by plotting the most common ages at which creative productions occur. By this method of compilation, much supportive evidence has emerged for the contention that intellectual creativity usually reaches its peak in early adulthood and therefore the conditions during these years ought to be made conducive to imaginative production. Lehman (1966) matched nearly 700 recognized scientific contributions since the year 1774 with the inventor's chronological age at the time of contribution. The peak period for deceased and living inventors was in the 30–40 age range.

The "time of plenty" view is not a pessimistic commentary. It simply points up certain facts about age and production. The reasons for these facts have retained their mystery though new hunches are continually being advanced and tested. On the one hand, the data would seem to urge young adults to put forth their best effort before reaching middle age. However, Lehman (1966) reminds us:

> Since group means tell nothing at all about individual performance, since each of the large number of some one individual's contributions may possess far more merit than the master works of certain other individuals, since there are numerous heartwarming exceptions to the general trend, and since, moreover, life should consist of far more than the production of scientific contributions, the present study provides no good reason why anyone should feel at any age that his usefulness is at an end. (p. 366)

Creativity can lead to higher profits in the business realm. This assertion is confirmed by companies like General Electric, IBM, and Kodak which point to innovations in procedure and product that arose from corporation sponsored training in creative thinking. Hoping to attain similar benefits, many other large and small companies have begun to orient their employees to techniques of generating creative ideas. Since the early 1980s creativity consultants have become a popular resource offering sessions for executives and production workers that range from one day in the cafeteria to week-long retreats in attractive settings (DeBono, 1984). Educators have taken notice of this demand for greater creativity. The National Institute of Education and state school boards in New York and Massachusetts are considering ways to incorporate creative thinking into the high school curriculums. Some universities with prestigious schools of business like Harvard, Stanford, and Michigan already offer courses on creativity to their students (Isakson & Parnes, 1985).

The Creative Thinking Process

Although the creative process is not fully understood, a number of phases are believed to commonly precede invention. These phases are termed preparation, incubation, illumination, and verification (Cagle, 1985).

Preparation. That preparation is a requisite for creative behavior often is a surprise to those who have supposed that "inspiration" just comes to some people and not to others. This assumption makes it easier to refrain from entering the great struggle in which all who create are engaged. After an initial self-prompting by vague insights, a creative person sets out to examine an area of difficulty by literally flooding himself with the diverse impressions of others.

The preparation stage is characterized by many dangers that can undermine production. First, the literature or information concerning the problem to be investigated may be so extensive or difficult that the task of preparation can seem overwhelming. At this point one may decide to drop the issue and move instead to another concern. A second and related danger is that side issues can capture attention and divert interest from the original purpose. This is especially common among those whose indiscriminate curiosity leads them in every direction (Isakson & Parnes, 1985).

Third, the very impatience that causes one to grapple with an issue can destroy all chance for success if one tries to begin synthesizing data prematurely, particularly given the presssures for rapid production. Again, the most important aspect of the preparation phase is immersion in the ideas of others. Awareness of how others have thought produces the material on which the synthesizing ability will operate. Getting out of the problem is easier at this stage than at any other, because little work has been invested and the degree of involvement is still small.

Many young adults often judge themselves to be uncreative because novel ideas haven't come to them without preparation. Reading about the real lives of eminent people—their failures, successes, and courage—can help correct this impression. Preparation is a difficult time for all creative persons. Those who underestimate its value will cease preparation too soon. Many potentially creative persons have given up because they have been led to believe that inventive thinking is easy or beyond their reach. In fact, very often what seems to be a creative achievement is basically a matter of abundant information.

Incubation. During the second phase of the creative process, incubation, there is an irrational, intuitive encounter with the subject matter. In this phase creative thinkers are afflicted with unrest as they try to arrive at the ordering structure, the recombination of data, that will merit their expression. They may be preoccupied to the point that they fail to attend to some of the most routine things expected of them. As they try to let the intuitive idea take conscious form, creative persons are dissatisfied with themselves and may be difficult to be around. Sometimes conflicts ensue with family members or close friends, who feel that the creative person's inattention to them is a deliberate insult.

In the incubation stage, self-doubt is an important mental-health hazard. Confidence is essential when unconscious activity brings up one thought combination after another. It is imperative that the conscious mind refrain from disapproving ideas as they emerge, deferring judgement until the entire unconscious product is available. Having to withhold judgment until the associative flow is terminated is an extremely difficult and demanding task that requires a high tolerance of frustration. Mental health is delicate during this phase of incubation, which may vary in duration from minutes to months. If the incubation period is protracted, one may lose important relationships with people who might have tolerated a bit of atypical behavior.

During incubation, anything disruptive to the focus of concentration is likely to be rejected. For some individuals incubation can take place on and off over a period of time, but for many creative persons the attempt to produce ideas leads to excessive measures for sustaining touch with the unconscious in a noisy and distracting world. Not everyone can spend the same amount of time in the tension-producing phase of incubation, but most creative persons prefer long blocks of time in which to work. This is without question the period when others must be understanding of the self-punishment.

Giving up at this stage is done only at great expense, for creative persons usually consider not achieving illumination as total failure (Guilford, 1979).

Illumination. If the incubation phase truly presents for most creative persons what Van Gogh called a "prison" in which they are confined by conversation and debate with themselves, then the illumination phase might be analogous to a release from jail with full pardon. Illumination is the inspirational moment the artist Paul Cézanne described as "a sort of liberation, the mysterious becoming external, everything falls into place—I see" (Ghiselin, 1952, p 16). It is the exhilarating triumph that creative persons like so much to relive, the time that is beyond words. Charles Darwin, whose search for the theory of evolution came to an end on a dusty lane, recalled "the very spot on the road whilst in my carriage, when to my joy the solution occurred to me" (Lowes, 1927, p. 443). Creative scientists, inventors, artists, and writers all look back in nostalgia at this brief but cherished moment and often speak of it as being mystical. With illumination, the burden of tension is lifted and creative persons regain touch with those around them.

Certain creative persons, especially those who have long awaited illumination, make an effort to retain its joy by sharing it. Usually their accounts of how an idea occurred are less than exciting to other people, who perceive the rapid transition from preoccupation and depression to joy and conscious delight as a further index of mental illness. They may wonder at the extremes in the creative person's behavior and especially at the sudden elation over something they consider unimportant. Further, in returning to a normal state of consciousness, some creative persons cannot figure out why people have become distant toward them during the interlude.

For some persons, the creative process ends with the illumination phase, for they have achieved a tentative answer and shed their tension. At this point, they may hasten to seek another problem. Persons of this inclination seldom attain recognition or contribute as much as they might, because they do not go on to make the form of their invention coherent to others (Guilford, 1979).

Verification. After an idea or a plan has emerged from unconscious activity, it must be consciously evaluated. Some creative persons find this need for verification difficult or even impossible to accept, because their emotional certainty about the worth of their idea or product precludes any criticism or adaptation. Nevertheless, the high pleasure of illumination must give way to rational judgment as the determinant of final production. If a writer is to communicate, the inspired work must be organized and edited. There must be a coherent flow, so that readers can share the message. Similarly, the successful experiment which elates the scientist must be clearly described.

Unlike the brief illumination phase, this verification phase can prove to be long, arduous, and at times disappointing to someone whose patience declines because of eagerness to undertake another quest. The hazard awaiting many writers, scientists, and creators in technology and art is the temptation not to follow through—a temptation that has prevented much good work from bearing fruit. There are many writers, like Coleridge and Shelley, who left fragments of unfinished work because they could not revise it, feeling that inspiration could not be improved upon and that any alteration would destroy illumination. Speaking of Gertrude Stein, Ernest Hemingway said, "She disliked the drudgery of revision and the obligation to make her writing intelligible" (1965, p. 176). On the other hand, certain writers have found that working out the implications of creative ideas can result in further discoveries. Hart Crane was an exacting author, and a look at his manuscripts indicates that his revisions entailed as much doubt as decision. Notwithstanding the attachment of some writers to their work, insight during the revision process can improve organization and structure.

Serious students of the creative process find it disappointing that the phase-sequence explanation includes only those events that occur after an individual has originally sensed a problem, become aware of some gap in knowledge, or experienced a vague insight into an area of difficulty. It is as though a reader who was interested in understanding a novel's elaborate design began reading at midpoint in the book. At the moment we do not know how much of the creative process actually precedes what are now described as its phases. It may be that our account of the creative process starts, not at the beginning, but close to the end.

Later Adulthood

Mental Functioning and Senility

Some older people become incapable of normal functioning. They forget things. They lose the theme of a conversation or even what they started to say. At times they go to another room or the store and cannot remember what they intended to do there. They become careless about health and grooming. Until recently, such manifestations in older persons have been attributed to senility. It was thought that the ineffective behaviors resulted from shrinking of the brain due to a nonreplacement of continuously dying cells. But brain cells, from the time of later childhood, do not (as do other cells) reproduce themselves. So adolescents, young, and middle-aged adults are also losing brain cells. Furthermore, most older persons do not suffer from senility. Despite the fact that they walk more slowly, their hair gets gray and thinner, they are less tempted to engage in vigorous exercise, that the skin becomes wrinkled and their facial lines deepen, some persons remain mentally alert, think accurately, and make wise decisions. This common circumstance is not just a function of good physical conditioning because many respected elder statesmen have delicate health or suffer physical handicaps. The brain and physical functioning are not strictly comparable. In fact, except for disease, accidents and certain health problems, senility is atypical and should not be considered a natural consequence of aging (Teri & Lewinson, 1985).

What then are the causes of senile dementia or mental confusion in old age? It is known that about 6 percent of the population over 65 suffers from marked mental decline from this affliction. In the past it was typically blamed on impaired blood circulation to the brain and consequent minor strokes. This view was generated by researchers who investigated institutional populations. Their subjects were readily available for testing and interviewing but probably not representative of the entire older age population. While stroke continues to be a cause, senility is now, in more than half of all cases, identified as Alzheimer's disease.

This degenerative and incurable brain disorder afflicts 4 million elderly Americans. The disease that mysteriously tangles nerve endings in the brain and ravages a victim's thought processes was named after Alois Alzheimer, the German physician who discovered it in 1906. During the course of the disease, which usually lasts from five to ten years, a patient's condition progressively deteriorates from memory loss to disorientation to complete helplessness and finally death. Alzheimer's is currently the fourth leading cause of death among the elderly, after heart disease, cancer, and stroke (Ennis, 1985).

Intensive research into the causes of Alzheimer's has been underway for less than a decade but some important clues have already emerged. It has been determined that the disease selectively destroys the brain area that produces acetylcholine, one of the chemicals or neurotransmitters responsible for carrying impulses between nerve cells. The chemical RNA has a key role in enabling the normal brain to synthesize protein continuously but Alzheimer's patients show a significant reduction in RNA and protein synthesis. Heredity is also implicated. Persons with a close relative who has the disease are four times more likely to get it. A genetic link is further corroborated by the fact that most cases of Down's syndrome—a form of mental retardation caused by an extra chromosome in the body's cells—develop what appears to be Alzheimer's after age 35. Families with an Alzheimer's member are three times more likely than others to also have a member with Down's syndrome. Consequently, scientists are working with the recombinant DNA to identify the gene that may be involved (*Newsweek*, 1984)

An examination of brain tissue after a patient dies is the only conclusive proof of Alzheimer's disease. But when the possibility is suspected, a less certain diagnosis by exclusion is used. Besides blood tests to ascertain whether anemia or thyroid might be causing the mental confusion, a CAT scan is used to detect signs of small strokes that can knock out areas of brain tissue. There are other treatable conditions like depression, alcoholism, malnutrition, and drug abuse that may produce symptoms which mimic Alzheimer's. The psychological workup includes tests of memory, attention span, language, spatial ability, and abstract reasoning.

When the judgement is reached that Alzheimer's is likely, treatment calls for keeping the patient mentally active, providing good nutrition and exercise. In order to offer the best possible home care, it is essential that the family be able to share its workload with adult day-care, or visiting nurse services. Somehow they must get away once or twice a week to go shopping, attend a movie—anything to keep things in perspective and maintain their own sanity. The Alzheimer's Disease and Related Disorder Association is a national support group that assembles families to share feelings of frustration and offers them guidance about handling their caretaking difficulties.

Intellectual Performance and Aging

A more cheering perspective on mental activity of the elderly than Alzheimer's disease is the actual norm. Baltes and Schaie (1974) were among the first to state "In our opinion general intellectual decline in old age is largely a myth... Our findings challenge the stereotyped view and promote a more optimistic one. We have discovered that the old man's boast 'I'm just as good as I ever was' may be true after all" (p. 35). In short, the loss of brain weight in adulthood is not synonymous with intellectual decline. Schaie, who directs the Gerontology Research Center at the University of Southern California conducted a longitudinal study of mental performance in adults. During this investigation, which covered a span of 21 years, several thousand healthy volunteers from 22 to 81 years of age were tested at seven-year intervals. It was found that, at all ages, the majority of the participants maintained or improved their level of intellectual competence as they grew older (National Institute on Aging, 1982).

There is no denying that aging does bring with it a physical slowing down and reduction of sheer strength, a loss of brain cell replacement and cumulative sensory deficit. What is not so often emphasized is that, should healthy older persons so choose, and if supported by others, there can be continued development of wisdom, compassion, generativity, empathy, common sense, and social perspective (Butler & Gleason, 1985).

Chapter 4

SUPPORTING EMOTIONAL GROWTH

Aspects of development such as cognition, morality, and socialization can be treated in a balanced way only when emotions are also considered. Love, affiliation, attachment, and negative feelings like anger, fear, and anxiety are parts of overall development. Emotions play a pivotal role in all aspects of growth, and in the nurturing of human behavior.

Positive, supportive, healthy emotions are products of intimate and caring interactions between persons of various ages. It is futile to try to separate the emotions of infants from the feelings of their caretakers. Indeed, the affective behavior of parents and surrogates form the basis for how infants react and what they will become emotionally.

The purpose of this chapter is to demonstrate that (1) emotional growth is a continuous lifelong process; (2) emotions are wholistic; i.e., inseparable from the mental, physical, and social aspects of human development; and (3) relationships between persons, young and old, are the building blocks of excitement or boredom, zest or indifference, fear or boldness, which accure from (4) interaction with fellow beings and influence our behaviors toward them.

THEORIES OF EMOTION

James-Lange Theory

William James and Carl Lange, concurrently but independently, formulated a theory that the physiological aspects of emotion were preceded by action (James, 1899). To clarify this description it might be compared to the common-sense perspective. Common sense says you see a bear, you are afraid, and you run. This is not the truth says James. The proper sequence is you see a bear, you run, and you become fearful. The response to the emotion evoking situation precedes the sensation of emotion. It is the response that stirs up adrenalin, stops digestion, raises blood pressure, causes facial flushing, and makes the mouth dry. It is not the sarcasm, the egocentric, unreasonable behavior of another person that provokes our anger. It is our own shouting and our counteraccusations that give rise to feelings of hostility and aggression. We can, in the face of someone's rudeness, egocentric and unreasonable behavior give a soft answer and minimize our own anger.

At mid-century catharsis was being explored. At that time it was supposed that children should be allowed freely to express their hostilities—"Let them get the bad things our of their system." Otherwise, boys and girls would suffer from their inhibition. Research studies demonstrated that allowing children to show aggression does not minimize its expression (Gaylin, 1984). On the contrary, according to James and Lange, it tends to aggravate chronic aggression. Most contemporary views of emotion are wholistic and emphasize a total involvement of the person; what one is, what one has learned, and what the situation is.

Thalamic Theory

The thalamic theory is sometimes called the Cannon theory, in honor of the first psychologist who called attention to the role of the thalamus and hypothalamus in the chemistry and expression of emotions. It was Cannon (1915), in his book *Bodily Changes in Pain, Hunger, Fear, and Rage* who gave us the basis for many current expressions of the "fight or flight" syndrome; i.e., the predisposition to run from, or attack, a stimulus.

The thalamus and hypothalamus are inner structures of the brain located at the top of the brain stem. It has been found that injuries to these areas have profound effects of emotional expression. Electrostimulation of the thalamus can cause the subject to report feelings, or sometimes manifestations, of violent or calm emotions. Such stimulation of certain sport in the thalamus area can, reportedly, cause fear; or have enough calming effect to stop the charge of an enraged bull. The theory involves the whole autonomic, or smooth-gland system (Malatesta & Izard, 1984).

Many attempts have been made to effect desirable emotional change by drug intervention. Seemingly, this is supported by, but was not sponsored by, the Cannon theory. Such intervention has not been successful. Certain drugs may calm disturbed persons sufficiently that they can better be treated by psychotherapy. Self-administered drugs to effect emotional change is opposed by physicians, clinical psychologists, and psychiatrists. We know, at high cost to the lives of creative, artistic, or athletic persons that emotions cannot be regulated just through direct treatment of autonomic, or thalamic structure. Certainly the last word is not in on therapy, the thalamus, and drugs. The use of medically prescribed drugs as a supplement to other intervention techniques is still being explored (Jenkins, 1985).

Activity Theory

This theory builds on the thalamic theory in postulating that the reticular system (network of fine nerves) must be active before the hypothalamus takes over its organizing and regulating functions. The reticular system is a network of cells located in the brain area occupied by the thalamus and hypothalamus. The person must be tense and excited if emotional response is to appear; otherwise, the person will remain calm and apathetic. If the subject has a damaged reticular system, of if that system has been anesthetized by drugs, emotional expression is inhibited.

The connection between the Cannon theory and the activity theory is acknowledged and explicitly stated (Lindsley, 1951). The arousal part of the activity theory has some similarity to the James-Lange position. Such relationships call attention to the fact that each theory emphasizes phenomena that seem previously to have been neglected or not sufficiently emphasized.

Cognitive Aspects of Emotion

Theories of emotion would be inadequately catalogued if they included only external stimulations, hormones (chemistry), and hypothalamus (autonomic controls) but omitted the role of cognition. Certainly we learn some, perhaps most, of our emotions. Persons do not normally behave as "primitive animals." Averill (1984) calls the learned aspect of emotions the social constructivist view. It assumes that emotional schemas are the internal representation of social norms or rules. It should be noted that this is more than a social and behavioral condition; it includes cognitive appraisal and internalization.

Emotions can be perceived as changes in organic state induced by stimulus situations. These learnings may, and often do, become enduring dispositions and/or temperaments. Thus, while it is recognized that endocrine glands and brain areas are operative in emotions, we can also acknowledge that there is individual input and choice. For example, the choice between panic and composure is partly a function of preplanning (e.g., women and children first) or training, as is the case of soldiers advancing under bombardment.

INFANCY

Developing Positive Emotions

Infants are credited with a large and rapidly growing repertoire of emotions. Such recognizable emotions as interest, surprise, disgust, and distress are present at birth. Anger, friendliness, sadness, and fear appear within a few weeks. The hope of some researchers is to help mothers interpret these first emotions more accurately so they can establish milieus conducive to the development of positive emotions.

The fact that infants have a wide range of emotions is significant for two reasons. First, it runs counter to the behaviorial and cognitive psychologies which have taught, for a long time, that emotions are learned through experience and conditioning. There is no denying that emotions are in part learned; but Bower (1982) contends those learnings are based on considerable congenital equipment. Second, if infants are so richly equipped with emotional readiness, a powerful argument for more attentive, loving, interactive infant care is presented to parents and their surrogates.

Bonding and Attachment

Current interest focuses on attachment and human bonding as the significant emotion of infancy. The age from about six to eighteen months has been found to be the critical period for developing the human response of love and attachment, and establishing the foundation for later socialization. Attachment is defined as an affectional tie that the baby forms for another person—mother or caretaker. This tie binds them together in space, time, and feeling (Bowlby, 1980).

Nonattachment and emotional deprivation are actually a devastating form of child abuse. A pioneer in this area of research, Fraiberg (1967), declared that the babies lacking close, intimate, caressing human contact during the first eighteen months of life develop the "tragic disease of non-attachment." They are robbed of their humanity. With high frequency these little a-humans can never become fully human because they can experience neither grief nor joy, neither guilt nor empathy. They cannot experience love and affection. Many, she says, as adults resort to violence in order to fan into life some sparks of feeling.

Lifelong Attitudinal Crises

There are, says Erikson (1963; 1980) critical, or sensitive periods of emotional development throughout life. He calls them normal identity crises, during which various aspects of personality may be formed. This psychoanalytic viewpoint employs the concept of polarity. The individual who fails the personality-growth challenge of a particular stage is at one pole (or extreme), and the person who is most successful is at the other. Of course, some people are neither highly successful nor utterly unsuccessful in completing a developmental task, so they retain fluctuating, unpredictable, ambiguous personality traits.

One of Erikson's ideas that has taken firm hold in psychological literature is that life's problems are never over. From birth to death we face tasks or crises of living. Success at one stage predisposes us to success in the next critical issue of personality development but does not guarantee it. Normal life crises overlap. So, emotional growth is continuous and gradual.

As age increases emotion becomes more subject to control by cognitive forces. It follows that there are differences in the emotional needs of infants and children and in the needs of adolescents and adults. The eight stages of emotional development or attitudinal crises postulated by Erikson are listed here and each will be discussed within the proper age context.

Approximate Ages	*Attitudinal Crises*
Birth–age 2	Basic Trust versus Mistrust
Ages 2–4	Autonomy versus Shame and Doubt
Ages 4–6	Initiative versus Guilt
Ages 6–12	Industry versus Inferiority
Ages 12–20	Identity versus Role Confusion
Ages 20–35	Intimacy versus Isolation
Ages 35–65	Generativity versus Stagnation
Age 65 +	Integrity versus Despair

Bear in mind that the *versus* in each stage reflects the psychoanalytic concept of polarity, i.e. ego versus id, pleasure versus guilt, conscious versus unconscious, love versus hate, and life versus death. According to Erikson, life consists in the struggle of the individual to lead a balanced life by opposing the negative forces (guilt, inferiority, selfishness) that would block out and diminish fulfillment of the positive, satisfying, social and personal urges.

Trust versus Mistrust

Life's first attitudinal crisis occurs between birth and age two. During this period an infant develops either a basic trust or a basic mistrust in things and people. Fortunate and healthy individuals learn to trust themselves, others, and the physical world. If infants are deprived of loving, nurturant contacts or subjected to excessive discontinuity in their care, they may experience a sense of loss and separation from the world and develop a chronic mistrust (Brazelton, 1983; White, 1985).

Theoretically, infants learn trust by having strong parents who provide tender-loving-care. The provision of this care depends more on the kind of persons caretakers are than on their childrearing techniques. This is a fact that has been repeatedly demonstrated. For example, abused babies learn fear and distrust; loved and cuddled babies learn trust and self-worth.

Early Childhood

Autonomy versus Shame and Doubt

In early childhood the potential for autonomy rests upon a child's motor skills. At two and three years of age youngsters are capable of climbing stairs, pushing and pulling objects, holding, letting go and throwing. These achievements invite an attitude of autonomy, of wanting to do certain things without any assistance from adults. Sometimes tasks like flushing the toilet, carrying a plate from the table to the sink, and getting dressed are handled well and result in feelings of self-control.

Of course there are also times when the caretakers become impatient because they have a schedule to meet. Then the usual procedure is either to rush the children, or do for them what they would be quite capable of doing themselves if there were sufficient time. If children are regularly exposed to protectiveness ("I'll do it for you—it takes you too long") or criticism for having accidents such as spilling the juice, dropping the plate or wetting pants, a sense of shame develops and the child comes to doubt whether he has the ability to control events and oneself.

Boys and girls need patient caretakers who will let them do some things on their own. The caretakers also help by providing structure, safety, and discipline. In order to experience pride and autonomy rather than feeling unworthy and lacking in confidence, children need to know limits, rules, and the necessity for routines. The world is too large, complex, and impersonal to face without continuity, safety rails, and "No trespassing" signs. However, children are ambivalently curious and cautious, eager and fearful, autonomous and dependent. Therefore they need caretakers who are emotionally responsive and accepting instead of punitive or verbally abusive. They need and want routine and order—a time to go to bed, a time to eat, a time to stop playing—and thus reduce threat of the unknown and unfamiliar by learning to live cooperatively (Erikson, 1980).

Handling Anger and Frustration

Anger and aggression are natural and, on occasion, lifesaving responses. The problem is learning how to handle these strong emotions in ways that are culturally appropriate to the circumstance. Anger is the child's way of saying that some situations are too difficult, that they cause frustration. Aggression is a form of anger in which the child attacks something or someone in retaliation. Such action is usually self-defeating because it results in counterattack or ostracism (Gaylin, 1984).

Before adopting techniques for helping children deal with anger and aggression, caretakers should try to understand what caused the emotional upset. This can best be done by observing the child and being a listener. For example, a sullen and sometimes aggressive boy was referred to a psychologist. He greeted the boy but received no response. The psychologist just went ahead and asked what the boy did yesterday. "I took a walk." Was anybody with you. "My dog." Do you talk to him? "Yes" What did you talk about? "My dad." This concerned questioning and the monosyllabic answers continued slowly. It turned out that the underlying factor in the boy's anger was being unable to fulfill unrealistic parental expectations. When the parents were so advised, they lightened up, looked for things to praise, and gradually the boy became more responsive, confident, and productive.

Getting children to talk about pictures they draw, talk to a friend on a play phone, tell what is happening to persons in a picture, or describe what stick figures are doing are other approaches to understanding. It is easier for children to do such things than it is for them to talk face to face with an adult who is much bigger than themselves.

Keeping in mind that the precipitating situation may not be the basic cause (as in the boy and his dog incident) some general techniques for enabling children to deal with anger and frustration are in order. Anderson (1983) suggests several guidelines:

> Dealing with anger and frustration should be motivated by a desire to understand rather than just to correct.
>
> Do not respond in anger. Punishment is ineffective if it is hostile.
>
> Set limits to the expression of anger but explain them.

Remove the misbehaving child from the situation but permit return when he can show self-control. Some children find this technique so effective that they can say to their preschool teacher "I guess I'd better leave the play area for a little while."

Let children know that you, the adult, see good in them.

Initiative versus Guilt: Feelings about school

Most four and five-year-olds are enrolled in some kind of group care setting when they face the emotional crisis Erikson (1980) refers to as initiative versus guilt. It teachers and parents respond to a child's many questions, they encourage a sense of curiosity or initiative for learning. Similarly, if they observe and participate in a child's creative play, they reinforce the value of initiative and imagination. By contrast, when caretakers cause children to feel that their questions are a nuisance and fantasy play is silly, then the normal initiatives of children are subordinated in favor of guilt.

One way to ensure that initiative predominates over guilt is for teachers and parents to collaborate. The experience of starting school can be difficult and typically occurs at an earlier age than in the past. Most teachers realize that the scope of their own observation is limited. Therefore, some of them rely upon parents to observe the child outside of school. When parents are asked to watch children at home, the resulting insights can contribute much to parents' and teachers' understanding about child adjustment. To illustrate, consider the larger view which emerges as parents join teachers in looking for indicators of success such as "Does your child like school?" Children usually find it less threatening to share with parents what they dislike about school. It follows that a teacher who assumes children like school unless they complain to her directly may overestimate student willingness to confide in her. At home, when parents hear a child's misgivings about the school experience, they can probe to find out "What did you tell the teacher about this problem?" The answer to such questions can reveal how comfortable the child is with expressing feelings in class.

Because a child's initial adjustment at school is considered so important by the teacher and parents, it is a good idea for them to share a common focus of observation, and then to talk frequently, at least once a month. Granted, this will require some improvisation of teacher scheduling and volunteer substitutes may be necessary. Nevertheless, the potential benefit to children more than justifies the accommodation by adults.

MIDDLE AND LATER CHILDHOOD

Industry versus Inferiority

The period from ages six to twelve results in an emergence of either industry or inferiority. As children become capable of deductive reasoning, they can participate in activities that require them to follow rules. Accordingly, boys and girls express considerable interest in the details of how things are made and what makes them work. This concern for industry is reflected by involvement with transformer toys that permit creative adaptation, constructing airplane models from a blueprint, planning and building forts and tree houses, engaging in craftwork, and cooking from a recipe. Sometimes these efforts lead to a product which adults approve or praise. In other cases, youngsters fail to follow directions or make mistakes which abort the desired result. When adults support additional experimentation, children are willing to try again and again. Soon youngsters realize that failure is part of the learning process. However, some grownups do no see it this way. Instead they are impatient and unduly critical, often leading children to believe they lack ability and are unlikely to experience success. In such cases boys and girls develop a sense of inferiority. Fortunately, even in these families children can still attain a sense of industry if their attempts to grow are supported by teachers at school.

Defining Success and Failure

Boys and girls typically look forward to entering elementary school. Under normal, ideal, and attainable circumstances their eager anticipation of life at school is confirmed. Children gain pride in themselves, their accomplishments, their expanding perspectives, and their new friends. Educational practices which foster positive emotions (self-confidence, friendliness, interest, initiative, trust, and industry) have been identified and described (Frymier, 1984). The policies needed are ones in which children experience a predominance of success over failure. All too often the barrier of interpersonal comparisons blot out the varieties and degrees of success individuals experience. The cloud of needing to be best, first, superior, above average makes many students feel that they are failures. Considerations that highlight individual success include:

> Students are accepted with their different skills (some can read before entering school, some have had little contact with books); some are socially competent with peers, some are just learning to get along with others; some are linguistically facile, some speak reluctantly.
> Teachers recognize the differences in motivation, aspiration, and skills that are due in part to varied socioeconomic status.
> This recognition of differences means that interpersonal comparisons are minimized or totally avoided.
> There is an equitable distribution of praise and recognition.
> There is equal access to privilege, opportunity and responsibility.
> Recognition of differences means that multiple teaching strategies will be used.

The matter of style of learning means contrasts such as working alone with a partner or group; learning best by listening, observing, doing, or reading; being independent or following teachers' directions; need for quiet or working with music.

Success must be determined by each child's progress from his status today to what it appropriately can be tomorrow or next week. Success, particularly in the early school years, should not be judged in terms of competition or comparison with others. In life as well as in the primary grades, there are multiple different living and learning styles.

Defining success/failure becomes part of an individual's way of defining oneself. The child is confident or lacks confidence. The adolescent believes he will succeed or fail. The adult defines self in terms of "I can/can't" perform well on the job. Defining success in terms of individual propensities can begin in the primary grades by stating varied aims of schooling: academic (mastery of fundamentals of learning); artistic and creative; leadership-followership; physical (health, skills in games); and communication (friendship).

Overcomers

Rutter (1984), a British psychiatrist, discovered that some children come through considerable adversity with confident and succeeding dispositions. They are overcomers. Rutter studied 90 well-adjusted, but ordinary, young women who had been put into institutions during their childhoods because of horrendous experiences. But at the institutions they reported satisfying experiences. "Most of the good experiences they reported were not academic success, but success in sports, responsibilities in school, or developing a good relationship with the teacher" (Rutter, 1984, p. 62). School authorities and child development specialists have been telling us for many years what Rutter has condensed into a sentence—"Parents and teachers should learn to help children define success in terms of unique potentialities."

Learning to Discuss feelings

Most of us are irritated by television reporters interviewing a person who has just experienced a tragedy or great victory. They thrust a microphone before the person and ask: "How do you feel about it?" Feelings are not readily discussed even under conditions of calm, i.e. in the presence of empathic persons and when there is time to be reflective. Discussion of one's feelings requires directed learning. However, when learned, discussion facilitates the chance for cognitive factors to magnify the impact of positive emotions and diminish the corrosive impact of negative emotions.

Preschool children have little trouble demonstrating, and are not reticent about verbalizing their feelings. "I hate you" or "I love you" is often a spontaneous comment. However, this candor is not always socially acceptable. Adults are apt to say, "You don't mean that." But, at the moment the child does mean it and has no doubt about the feelings. Controlling the statement does not necessarily help. The containment of anger permits it to remain as an inner poison that further reduces the zest for overcoming the real and external cause. Repression also allows for the one-sided, illogical thoughts that compound one's own deficiencies and magnifies another's culpability.

Talking about one's feelings provides an avenue for catharsis—getting it off one's chest and for achieving balanced perspective.

The matter of expression/repression is a lifelong dilemma and childhood is a good time to begin learning to discuss feelings. Hence, in addition to saying "You just don't say such things," one can add, "...at the moment. Let's cool off a bit and then talk about it." These suggestions can improve dialogue with children:

> Respect the feelings the person has expressed. (No admonition, "You should not feel that way.")
>
> Listen carefully (wait your turn rather than interrupt with words of wisdom).
>
> Reflect the other person's comments (i.e. rephrase what you think you have just heard).
>
> Speak as though you were conversing with an adult.
>
> Ask his opinion when appropriate (the idea is to make your verbal partner feel significant).
>
> Help the child feel good about himself. Bear in mind that the ultimate purpose of the dialogue is to enhance the child's self-esteem.

All of these recommendations will be easier if dialogue with children is a routine part of family and school living.

Dealing with anger and aggression goes further than just controlling these emotions. Development includes increases in self-gratification and altruism. Positive emotions (interest, self-confidence, affiliation) facilitate cognitive performance. It takes emotionally mature caretakers to guide children into healthy emotional discussion and development. Teaching children to discuss emotions is not just something adults do; it is also a matter of the kind of persons the adults are.

Sharing Fears

To clarify the role parents and teachers should assume in helping children learn to cope with fear, we need to examine our own response to the irrational. It is well known that boys and girls live more with fear than do adults. This difference can be explained in part by the greater access children have to imagination. For them, the day includes more fiction, and sleep brings more nightmares than grownups experience. Nevertheless, many parents and relatives are ashamed of fearful children and attempt to banish their fright by denial. For example, the author took his kindergarten son to the zoo. Since Steven was fascinated by crocodiles, we spent quite some time at the reptile exhibit. While we were in the aquarium, another little boy arrived with his family. The youngster was apprehensive and preferred to stay at some distance from the floor-to-ceiling window behind which the crocodiles lay. Taking notice of his fear, the boy's father lifted him up, then held him against the window and announced "See, it's like I told you before; they're locked in, so you have no reason to be afraid." But the young child continued screaming with fear. Imagine how confusing it is for a little boy to be told he is not afraid when he really is. His parents, those all-knowing authorities, must know more than he does. Some children thus develop an alienation from their own feelings. They learn to mistrust their senses and rely on other people to tell them what to feel (Goleman, 1985).

Ridicule is also a common response to fear. But laughing at another's fears does not decrease the fear. Instead the effect is to lower the individual's self-confidence. Statements like "There is nothing to be afraid of in the dark," or "It's just a dream and not real" may be well intended; but they still inspire shame for having fears that grownups declare to be unwarranted. To laugh at someone's fears or to call a person a sissy or a baby is to limit the relationship. Children and adults whose feelings are ridiculed soon stop sharing their experiences. The tragedy for parents is that they reduce the chance to know their child better and forfeit an opportunity to help the youngster learn to cope with doubts and fears.

The first step in overcoming fear is to acknowledge it. Because young children identify closely with parents, a sound way for parents to reduce the harmful consequences of children's fears is to admit their own fears. A child who is afraid of the dark, of being alone at home, of starting school, of being intimidated by bullies should be assured that fear is a natural reaction and that admitting fear does not make one a coward or sissy. Although parents may insist there is no danger in the dark, a child knows that Mom chooses not to go out alone late at night and that the doors are locked at night. A child should not be made to feel ashamed of expressing feelings of fear. After all, adults are afraid of many things too—getting old, being rejected, losing their hair, developing cancer, taking tests, and being alone. However, adults have an idea of what fear is, whereas children are afraid long before they can comprehend the notion of fear (Allman, 1985).

The child whose fears are unsuspected or unshared by loved ones bears the added burden of loneliness. Certainly children should know someone who will hear their doubts and fears without judging them. They will have this experience if parents accept them unconditionally. It should be unnecessary for sons and daughters to repress fear or feign bravery to gain their parents' esteem. The way in which mother and father handle their own feelings is also important. Because youngsters have a strong tendency to emulate their parents' behavior, parents can, by their example, lead children to express feelings honestly or learn to deny them.

Drugs and Anxiety

Most users of drugs and alcohol do not begin their involvement until high school. Many of them can avoid participation if they are exposed to prevention programs in elementary school. Fortunately, this is happening in cities throughout the country. These efforts will need to continue and grow or we face the prospect of countless ruined lives. To illustrate, it is estimated that half of the alcohol-related highway fatalities involve teenage drivers. Consequently, a number of states have raised the legal drinking age. Widespread education programs are also underway in high school to reduce the incidence of drug-taking.

Despite these measures the scope of chemical abuse seems destined to rise because of a frightening new phenomenon called designer drugs. Theses are man-made imitations of heroin and other substances. Because synthetics are cheap to make they can be sold at a low cost. To make matters worse, the clandestine chemists who invent these drugs have no means of quality control, no way to know whether they are producing killer batches. Chemical analyses by the Federal Drug Enforcement Agency reveals that in some cases the synthetics can be up to 3000 times as strong as the real thing. These drugs have not been tested on animals so the human addict population is providing our first look at the possible consequences. For example, certain of the drugs have been found to induce instant, incurable Parkinson's disease (Elliott, Huizinga, & Ageton, 1985).

The need for early consideration about the hazards of alcohol and drugs certainly seems warranted. If one doubts the wisdom of group discussion by preadolescents on the topic of drug abuse or experimentation, note the remarks of an increasing number of students: "Users turn me off. They're jerks." "They're spaced out in class. That's not for me." "I want more out of life." "They talk slowly and stare a lot. It messes up their minds."

The development tasks of later childhood include achieving personal independence, building wholesome attitudes toward self, forming constructive relationships with peers, and making a start toward clarifying and demonstrating one's values. Group discussion can make contributions to all of these tasks. Consequently, a challenge for adults is to provide a place and setting for such discussion. Because of the first task cited (gaining independence) the adult facilitator must avoid leading the group and overwhelming it with facts, knowledge, personal experience, advice, and moralizing. Simply make sure that when wise words are uttered by the preadolescents (e.g. "Marijuana messes up their minds") the words are emphasized by a nod or "What did you say?" or "Say that again, please." The facilitator can highlight the second cited developmental task (attitudes toward self) by finding some way to repeat and emphasize the focal point in the National Institute on Drug Abuse campaign, "Just Say No." It is the individual student who makes the declaration. "Just Say No" can be made the emphasis when the discussion deals with peer relations and developing a scale of values.

Adolescence

Identity versus Role Confusion

In adolescence a new perspective of self and others is brought on by physical and mental changes. In addition to looking more like adults, teenagers also become able to conceive of ideal environments or relationships and then contrast these with their own circumstance. For some the result is discontent; others are motivated to bring about constructive change. The sense of idealism enables teenagers to picture how the world and its often conflicting elements might be brought together in peaceful coexistence.

The inclination at this stage is to combine the views a person has acquired about the self in various roles as a son or daughter, student, friend, athlete and member of peer groups. This task is facilitated by the newfound integrative abilities which emerge during adolescence. When a person is able to bring these images together, recognize their continuity, and derive assurance, he can look to the future with confidence. These persons have a sense of identity because they know who they are, where they come from and where they are going.

Erikson (1980) believes adolecents can either establish a more or less firm identity or become lost in role diffusion. If they achieve identity, they can integrate their childhood experiences with new biological drives, expanding abilities, and the need to perform a social role. If the task is too great and if perceptive help is not provided, adolescents escape by establishing a negative identity; that is, they become what parents and society do not want them to become. They lack roots.

Being without a role and lacking a sense of identity can have serious consequences. In the extreme a person many find this condition so abhorrent as to choose self-destruction. Presently suicide is the third major cause of death among teenagers. This age group accounts for more than 6,000 of the nation's annual 25,000 suicides (Allman, 1985). Conflict with parents, school difficulties, family dissolution, facing an uncertain future, broken friendships, loss of a loved one, peer rejection—any of these and other factors can undermine a person's sense of identity and cause him to give up, to seek the ultimate escape. Although the expansion of suicide prevention centers is a favorable response, it is clearly insufficient. One estimate is that 98 percent of the persons who commit suicide never contact these resources for help (Santrock & Yussen, 1984).

Maintaining Self-Concept: The Role of Defense Mechanisms

More than anyone else, Sigmund Freud influenced modern thinking about the internal processes of self-assessment. He identified the defense mechanisms and showed how we rely on them throughout life to maintain a favorable self-impression. To fully appreciate the way in which our defense mechanisms operate, it is necessary to understand something of Freud's overall thesis regarding emotional control.

Freud (1925; 1962) contented that much of life is shaped by the urgings that derive from the unconscious. Within the irrational unconscious there is a striving for gratification, a pleasure-bound inclination Freud called the id. The id knows no values, no right or wrong, no moral standard. Obviously life could not proceed if every person were to behave solely in response to the unconscious strivings of the id. There would be rampant theft, murder, and rape. Fortunately, there is a censor mechanism acting to restrain the id's primitive desires and irrational demands. Freud called this mechanism the ego. The ego's goal is self-preservation. In contrast to the unconscious id, the conscious ego is capable of calculating the consequences of behavior. Whereas the id does not care for life that does not offer pleasure, the ego does not care for pleasure that jeopardizes life. When the irrational id suggests that a person go ahead with an unreasonable but pleasurable activity, the rational ego (sometimes prompted by the superego, or moral conscience) responds by presenting the self with reasons why it ought not to proceed in this way at this time. Each of us regularly experience inner dialogue while talking to ourselves about possible courses of action.

In order to defend against the id's desires—which, if continually satisfied would destroy a person—the ego must enable the person to sustain a high level of self-esteem. Otherwise desire would bypass the censor (ego) and lead to self-derogation and destruction. Sometimes, in order to think well of ourselves when circumstances do not warrant it, the ego must distort reality, changing the facts so that we see our behavior as commendable. Freud called the various ways by which the ego distorts reality defense mechanisms (Goleman, 1985).

All of the defense mechanisms represent attempts to maintain a positive self-concept. Familiar examples of reality distortion include repression, daydreaming, projection, rationalization. By repression is meant the common tendency to force unpleasant experiences from the realm of the conscious, to bury them so they can no longer be a bother. Projection is the mechanism by which unaccepted aspects of our own behavior are attributed to other people; that is, we condemn in other people those characteristics of ourselves we cannot accept. In daydreaming the ego protects us so that we never find ourselves in trouble, as is sometimes the case in night dreams. In daydreams we always win the game, get the girl, become the movie star, turn out to be a heroine, or achieve our academic ambitions. Rationalization means the opposite of what the word rational implies. It is the irrational, illogical, evasive assigning of false reasons for failure, inadequacy, or error. For example, a boy who gets a failing mark on an assigned paper might defend himself (to himself) by maintaining that all his classmates cheated, but he kept his integrity. Or, he might try to convince himself that, because he was the only boy in class, his poor grade reflects the teacher's bias in favor of girls.

It is clear that, for most of us, the defense mechanisms perform a valuable self-preservation function by maintaining our access to a favorable self-concept. Teachers should recognize that all students occasionally need to rely on defense mechanisms and that certain defenses are more common at specific ages than at others. For example, during junior high school, denial is a common response to the teacher's question: "Now, who doesn't understand the algebra homework?" Even though most of the class may not comprehend how to do the assignment, they do not wish to appear stupid by admitting the fact in front of peers. The frequency of daydreaming is highest during junior and senior high school when students are preoccupied with pimples and other obstacles to the attractive physical appearance they desire. Rationalization is common beginning in adolescence.

Teachers can behave in ways that minimize students' need to use defense mechanisms. Specifically, teachers can: encourage and respect students' questions and ideas, enabling them to recognize the worth of their own thinking; encourage the use of unique strengths; help groups establish standards in which members can take pride; keep personal information confidential; let students save face in all situations; and explicitly recognize improved performance. To avoid the predictable reliance on defenses that occur when failure is continual, teachers can help students detect the causes of their mistakes, and remediate inadequate performance. Further, they can assist students to develop the abilities to accept constructive criticism and learn to appraise themselves realistically.

Peer Relations and Locus of Control

As adolescents widen their social world they move away from dependence on parents and search for congenial peers. Parents may be hesitant about accepting the new sets of friends for fear they will be the "wrong kind," i.e. morally different, drug experimenters, shoplifters, sexually permissive, troublemakers, etc. The fear is well founded. Recent years have seen widespread drug abuse, increasing sexual promiscuity, more venereal disease and growing numbers of babies born to teenage girls. In contrast to the seriousness of the move away from home is the threat of youth who have remained attached, tied to mother's apron strings. The assumption of increasing responsibility and gradually moving away from home must ultimately be regarded as a step in the right direction. It is a move in the direction of desirable emotional and social autonomy.

A convenient way to summarize the challenges inherent in developing peer relationships and loosening of parental ties is the concept of locus of control. This relates to the feelings one has about the forces that shape one's life. If one believes that fate, luck, destiny, and other people determine success, failure, aspirations, and actions, one is said to have an external locus of control (ELOC). Those who believe that personal actions, individual perspective, persistence and optimism chart their lifestyles, success, and future have an internal locus of control (ILOC).

Adolescents' locus of control influences the way they apply themselves to school tasks, their attitudes toward grades, their responsiveness to teacher dominance or democratic procedures, their susceptibility to crowd behavior, and the stance they take toward the assumption of responsibility. It influences the zest they have for self-actualization of their own potential. Bradley and Teeter (1977) reported that ILOC students tended to be considerate of others, task-oriented, and persistent. ELOC students were more likely to be distractable, inclined to hostility, and prejudiced.

It has often been suggested that membership in the lower socioeconomic class tends to shape an ELOC; and that upper middle class students more frequently develop an ILOC. Eron and Peterson (1982), collating many studies, emphasized that the lifestyle of parents and the instructional style of teachers were more influential in shaping LOC than socioeconomic status. That is, parents who themselves—be they black, Hispanic or white, middle or lower class—were independent, self-directed and optimistic tended to have ILOC offspring. Childrearing methods rather than educational level, are related to LOC. Teaching that effects skill development, self-direction, self-confidence, and individuality are similarly conducive to ILOC.

Early Adulthood

Theories of Adult Emotion

The constructivist theory of emotions emphasizes the social rather than biological factors that provide the genesis for adult emotions. Culture and social roles, which are flexibly open-ended, provide the basis for acquiring new and more mature emotions. This impression is both challenging and discouraging. It is challenging because adults can be active factors in their own behavior. They do not have to remain the victims of parental abuse, economic or cultural deprivation, parental separation, or other stultifying childhood experiences. It is discouraging because it tempers the defense mechanisms adults so readily use to explain their failure to face their own feelings and inadequacies. It magnifies personal accountability. Averill (1984) suggests that mobility, job changes, domestic responsibilities or failures might provide reasons for facing one's own feelings. Meditation, counseling, religious involvement, or group discussion may provide the medium for getting in touch with one's true feelings and directing the course of further emotional development.

Roseman's (1984) structural theory emphasizes the factor of individual input for emotional development during adulthood. He contends there are cognitive determinants in both the generation and expression of emotions. Organic and experience factors are not ignored but it is advisable also to acknowledge the cognitive factor. Emotions have both a phenomenological component (thoughts, feelings, and images) and also physiological components (neural, hormonal, and visceral).

Intimacy versus Isolation

The developmental tasks of early adulthood include getting started in an occupation, finding a congenial social group, selecting a mate, learning to live with a marriage partner, starting a family, raising children, managing a home, and taking on civic responsibility. At the same time a person assumes these difficult roles and obligations one is also expected to meet the emotional challenge of achieving intimacy. For those who have established a sense of identity, this is a manageable goal. Erikson (1980) cites such examples of intimacy as reliable friendships, concern for the less fortunate, the willingness to lose some of one's privacy and eagerness to fuse one's life with a mate. In marriage it means sharing work, dreams, leisure, and procreation with another person and securing satisfactory care for the offspring.

Compassion and empathy are moods that approximate this fusing of lives. Ego or self concerns diminish with maturity or are so thoroughly merged with the lives of others that they are indistinguishable, i.e. one gets personal satisfaction from having a part in the satisfaction of others. Developing these motives seems to come easily for certain people. The beginnings may be seen even in later childhood when some boys and girls agonize over the pain of pets, siblings, and friends. In contrast, it is a struggle for some adults to move away from immature, identity-seeking, competitive jousting and self-centered pursuits. Bruner (1982, p. 58) asserts that he does not see how a society can be run "... without some tacit recognition of the importance of compassion—of helping those less fortunate or even less able than yourself." In both the family and community, those who do not achieve an open orientation toward others feel a sense of isolation and alienation. Successful efforts at this time in life are an outgrowth of earlier stages, but the development of social maturity (intimacy) is the basic crisis during early adulthood.

Enjoying Parenthood

It is customary for discussions about parenting to center on its demands and disappointments. There are frequent reports about child abuse, runaway parents or children, problems of communication with teenagers, and how to handle the aftermath of divorce. All of these concerns deserve attention. But it is also helpful to view the brighter side, to learn from satisfied parents what they enjoy about their role.

This was the focus of an interdisciplinary research group formed to help couples determine whether they should have children (Peck & Granzig, 1978). Before a parenthood aptitude test could be devised, it was necessary to examine the experience of happy parents. Nearly 1,000 mothers and fathers were included in the profile group based on these criteria:

> Because adolescents are regarded as the most difficult age group to take care of, all of the successful parents had to have at least one teenager.
>
> Parents' success had to be confirmed by two professionals from differing fields; a teacher for each child, a relative from each side of the family, and by their own child(ren).

The selected parents emphasized emotional satisfactions when describing what they liked about their role. For example, the joys of interacting with sons and daughters, sharing special moments and getting to know the children were mentioned frequently. Most families did not identify a child's special talent or school achievement as having an influence on their relationship. Instead the parents rooted their pride in the child's personality. When asked to advise other less satisfied parents about how to enjoy children more, the following recommendations were made:

(1) *Be ready to become a parent.* In retrospect many successful mothers and fathers felt that their decision at the time of marriage to delay having children was a good idea. In some cases this choice allowed women to pursue career aspirations. Later, the children were not viewed with resentment or seen as having interfered with an occupational goal. Some families felt that being able to save some money, buy their own home, and become more mature before assuming parenthood made it easier for them to devote greater time to the children when they arrived. It seems that couples do a better job of parenting when they feel ready for the task. This conclusion is supported by childrearing experts who believe readiness to be a vital factor in predicting parental success. Many university students are struggling but unhappy as they try to adjust to all the demands a new baby presents. By waiting several years to become parents, preferably at about age 25, they would find their childrearing role more pleasing. The benefits of waiting to start a family are underscored by U.S. Census Bureau data showing a decreased risk of divorce with increased length of marriage before a couple has children. This suggests that maturity of a couple's relationship may be as important as their individual maturity in determining readiness for parenthood (Feldman & Feldman, 1985).

(2) *Appreciate the changing nature of the parent-child relationship.* Successful parents have been able to discover continuing satisfactions at every child age level and thus maintain a consistently positive view of their changing role. They enjoyed nurturing the infant, assisting the preschooler with fundamentals of learning, helping with peer problems and teaching leisure skills during the elementary grades and, still later, discussing with teenagers choices the adolescent will make on their own. By taking seriously the need for change and growth in themselves as a factor in fostering their child's development, these kind of parents are able to sustain a long-term mutually satisfying relationship with sons and daughters. Other studies also indicate that maturity is the most distinctive characteristic of healthy families (Green & Kolevzon, 1984).

(3) *Give your children the priority they deserve.* Every mother and father has to decide how much time will be spent with their children.

Peck & Leiberman (1981) note that parents in the success sample differed from other less satisfied parents in the clear priority they assigned to their family. Certainly they did not devote themselves exclusively to raising their children; but whenever a conflict arose between home and job, there was no doubt about the choice. One mother stated the case for her peers: "When my daughter was born I kept working for a while. Then I realized I didn't want to spend the extra hours at the office that would have led to a partnership for me. My daughter is only going to have one childhood, and I don't want to miss it." Another successful mother described her feelings: "I'd like to tell every young woman who does not yet have children that it is nonsense to think that having a child doesn't have to interfere with a career—it should."

Parents differ in the value they assign to conversations with young children. Some of them view the exchange as a delight, while others regard it as an unwelcome distraction. Later, however, when the same children reach adolescence, some parents reverse themselves by complaining that they cannot get their teenagers to talk to them and confide in them. Usually they explain this loss in communication as though it represents normal development, a natural change in the teenage years indicating greater reliance on peers and the quest for independence. The fact that some parents—like those in the success sample—continue to enjoy access to the feelings of sons and daughters once they become adolescents and continue to be sought after for advice is often interpreted as a stroke of luck. The truth is that at every age we share ourselves most with whomever has demonstrated willingness to spend time with us.

MIDDLE AGE

Generativity versus Stagnation

When men and women reach middle age, personality development calls for a choice between generativity and stagnation. The concept of generativity means enlarging one's concern for others, going beyond the immediate family to care about unknown people who live in other places and perhaps less fortunate circumstances. A sense of generativity also creates concern about people who are not yet alive, the future generations whose welfare will be influenced by how the present population manages natural resources, protects and preserves the environment, and works to avoid nuclear war. This mature outlook can be observed in unmarried as well as married persons; it is increasingly seen in the behavior of persons who are younger than middle-aged. Probably this earlier show of maturity is because satellite and cable television are creating an unprecedented worldwide kinship. Charitable efforts such as the Live-Aid global music concert in behalf of starving people in Africa inspires generativity among all age groups. Anyone who feels a commitment to mankind, who reaches out and tries to make life better for others through volunteerism, civic involvement, or financial giving demonstrates generativity. In midlife these individuals recognize the transience of life and want to leave a worthwhile legacy, to have contributed in some way.

An opposite view obtains among those who decide to live only for the moment—for the present—by indulging themselves in comforts and regarding personal desires as always deserving their highest priority. Such persons fall into a condition Erikson (1980) calls stagnation. Scrooge, the insensitive and money-possessed character in Dickens' *A Christmas Carol* epitomizes stagnation. Like many others before and since his time, Scrooge eventually decides that intimacy and generativity make life more meaningful and support mental health.

Long-lasting Marriages

The average American marriage lasts for nine years. More than a million couples annually are granted a divorce. Their disappointing experiences are frequently shared on television talk shows, in advice columns, magazine articles, books and self-help discussion groups. So extensive is the coverage of broken relationships that the general public sometimes feels overinformed on the subject of how marriages fail. It would seem that to raise the marriage survival rate some attention should be paid to the experience of couples who have managed to remain together for many years.

This was the rationale of Lauer and Lauer (1985) at the International University in San Diego. With the help of colleagues and students they identified 351 couples who had been married for 15 years or more. Most of the couples, 300 of them, claimed to be happily married, a few (19) were staying together for other reasons such as the children, and some (32) consisted of only one unhappy partner. All of the participants individually completed a questionnaire which gave a profile of their martial relationship. Then each person was asked to choose from his or her answers the ones which best described why their marriage had been sustained. Consider some of the most frequently given reasons.

(1) *My spouse is my best friend.* The reason most often given for a long-term relationship was friendship. Liking one's partner, as well as loving them, means enjoying each other's company and feeling satisfaction in spending time together. Many of the happily married husbands and wives reported that their mates grew more fascinating with the passage of time. "She's more precious to me now than when we were first going together." There is a special joy in watching a partner become more mature and share the benefits in terms of greater intimacy. The qualities couples most admired in their spouse were caring, giving, integrity, and a sense of humor. The consistency of these behaviors made a person's faults easier to live with and more amenable to change.

(2) *Marriage is sacred and a lifelong commitment.* The common vow of "till death do us part" was taken seriously by these couples as a reasonable expectation for their marriage duration. Some men and women mentioned that no one can avoid troubled times so commitment requires a willingness to be unhappy once in a while. One may not always by pleased with the behavior of one's spouse. But, instead of walking away from the marriage, both parties must learn together how to resolve problems. Most of the participants did not consider divorce as an option in dealing with family strife. In fact, the group was quite critical of divorced peers who they felt seem to underestimate what it takes to make a marriage work. This assertion is confirmed to some degree by those who experience divorce. Packard (1983) found that two years after their decree, 60 percent of men and 72 percent of women thought their divorce might have been a mistake and that they should have tried harder to resolve their problems.

A national Gallup (1985a) poll of teenagers' perceptions also deserves mention. In response to the question, "Generally speaking, do you think it is too easy or not easy enough for people in this country to get divorced?" Seventy-five percent of the adolescents believe divorce is too easy to obtain. That figure has risen from 55 percent when the question was initially posed in 1978. The teenagers were also invited to think about this issue: "Generally speaking, do you think that most people who get divorces have tried hard enough to save their marriage or not?" A similar proportion, 74 percent believe that divorced couples did not try hard enough to save their marriage (a rise from 62 percent in 1978).

(3) *We agree on aims and goals*. Besides the attitudes they share regarding marriage and their mutual perception of each other as the best friend, successful spouses also agree about the purposes and goals in their lives. Sharing commitments and beliefs does not mean there are no differences of opinion. Both parties have learned to discuss differences and maintain a sense of humor as part of the process of reaching agreements. Unexpectedly, fewer than ten percent of the spouses identified sex as a factor in keeping them together. This is not an index of dissatisfaction. Actually 70 percent expressed great satisfaction with the intimate aspect of their life. But they did not feel that sex is a sufficient basis on which to build a lasting relationship. For this reason and because of strong commitments to each other, infidelity was an uncommon problem.

(4) *We resolve differences*. There were other unexpected patterns of behavior among the couples. For example, they opposed the free expression of anger during family conflict. Rather, their individual practice was to remove themselves from a discussion when they got mad. Later, they would resume the interaction and make comments without shouting. These reflective remarks along with respectful listening habits were thought to be a better method of resolving disputes than personal assertion and shouting. Another surprise involved attitudes toward equality within the family. Most couples felt that the notion of a 50-50 division of responsibility and effort is unreasonable. Instead they believe that each person should be willing to give more than she or he receives most of the time. In the long run the proportion of giving and taking will even out. When either person enters a marriage with the expectation that everything must be equal they're headed for trouble. One husband said: "I sometimes give much more than I receive and in other circumstances, it is the opposite. If my wife and I were unwilling to take this view, we would probably have parted long ago."

(5) *We respect identity and individuality*. Spouses should pursue some separate interests in order to maintain their own identities. This is the usual advice of marriage counselors. But the couples in this study disagree. Most of them make an effort to spend as much time together as possible. The attitude is "She is my best friend. I would rather speak with her and do things with her than anyone else." Some who work together full-time still felt they had too few moments to share. Lauer and Lauer (1985) did not observe any loss of individuality in those who seemed to enjoy each other's constant company.

Surmounting Loneliness

Persons at any age-stage can be lonely—witness the autistic child or the incapacitated elderly man confined largely to one room and without visitors other than institutional custodians. It seems appropriate to discuss this widespread, desperation-producing malady in connection with middle age for several reasons: Grown children have left home and nonemployed mothers may feel their lives are now lacking purpose. Men are reaching the peak of their vocational careers and may feel that they are being set aside, passed over, and ignored by younger coworkers. Changes in sexual desire are too often seen, not as changes in self but as evidence that one's spouse cares less and is less romantic. Lacking the vigor to zestfully pursue home duties, vocational pressures, and strenuous leisure the range of one's activities decreases. Persons tend to sit at home watching television, sipping a drink, and feeling lonely (Eisenberg & Eisenberg, 1985).

Loneliness, defined as being alone, feeling separated, isolated, and longing for friends, is aggravated by such things as technological society (machines replacing workers), changes in sexual mores (waning respect for marriage), substitution of fantasy and mass entertainment for involvement (x-rated videos), and diminished strength of the extended family. The population explosion with its accompanying anonymity of individuals also contribute to loneliness.

Rook (1984) makes the point that loneliness is a subjective condition that is acknowledged and so labeled by an individual. This would seem to place major responsibility on the person to change his or her thoughts and actions. Rook does not offer such a simplistic solution. She emphasizes the fact of social support that is needed by humans (who are inherently social creatures). This support consists in *emotional support*—esteem, trust, concern, listening; *appraisal support*—affirmation, constructive feedback; *informational support*—suggestions, information, advice, direction; and, *instrumental support*—money, labor, time, instruction, organized action.

LATER ADULTHOOD

Contrasts Characterize Later Life

The stereotype of emotions in old age is one of loneliness, apathy, futility, and confusion. Persons who portray all senior adults in this way engage in ageism. As with racism, ageism leads to prejudicial thought and behavior. And, like other victims of prejudice, it may cause old persons to believe derogatory things about themselves. This is attested to by the considerable number of people who move toward later life with dread and anxiety. But there is another picture of men and women who approach retirement with optimistic anticipation. They eagerly look forward to traveling to places they have dreamed about. Some retirees expect to exercise latent talents they have not previously been able to use. Butler and Gleason (1985) report the positive emotional and personality shifts that can accompany aging. The consensus is that the changes are highly individual, not general.

Because of experience, older persons have an improved chance for wisely choosing the course of their further emotional development. However, younger people should understand that although new circumstances occur, the accretion of past habits tend to keep persons from making abrupt changes. The complainers of early and middle adulthood find it more difficult to be confident, optimistic goal-seekers in later life. On the positive side, those who were caring, responsible, striving young people will likely continue these good habits when they reach their sixty-fifth birthday. Whatever age one is—adolescent, young adult, middle life—the time is now for building emotional maturity in old age.

Ego Integrity versus Despair

Erikson (1980) postulates that the developmental crisis of old age is integrity versus despair. If integrity prevails elderly persons believe they have taken care of themselves, their affairs, and their responsibilities competently. They have adapted to the triumphs and disappointments of life. They have accepted the decline of their careers, incomes, youthful vigor, and beauty. They have achieved emotional integration with the stream of mankind. They are, says Erikson, prepared to take the final step in the life cycle: death.

If, on the other hand, the feelings of despair prevail, elderly persons conclude that their lives have been futile, a series of missed opportunities; they feel they are detached from people (including their children and grandchildren). Any purpose they had in life has been obliterated by retirement. They are ill-prepared for death. They suffer remorse for not having succeeded and daydream about prolonged life and having another chance.

Attitudes Toward Death and Dying

Few persons face the prospect of immediate death with eager anticipation—the exceptions are those suffering pain from illness, injury, or psychological trauma. Otherwise, the emotions involved in facing immediate death range from willingness, through acceptance, dread, anger, and fear. Fortunately, healthy emotional development throughout life brings an increased readiness and acceptance of death and a reduction in fear and anger. This increased acceptance is fortunate because the death rate per thousand rises dramatically for those who are over age sixty-five.

Acceptance of death and decrease in fear and anger is needed because death is both inevitable and, over the lifespan, unpredictable; i.e., it occurs in all age-stages. There is a growing belief in the helping professions that death should be talked about by young and old more frequently than has been the case (Kastenbaum, 1986). An increasing number of college have offered, in the past two decades, courses and seminars on death and dying. There is only slight evidence that such courses have reduced substantially the discomfort and inability to face death. There is, however, a firm belief that such study does no harm. But courses are not enough. Before there can be a more ready acceptance of death there must be strong ties that replace the rootlessness that now characterizes our society. In short, the goals for optimum human development also function in meeting the challenge for confronting death with equanimity.

Just as adolescence is a cultural creation, so too our attitude toward death is a cultural invention. This observation is well documented by recent medical developments that make it difficult even to define death. Mechanical hearts, mechanical lungs, mechanical feeding and mechanical elimination have made it hard to say just when a person is dead. This clinical uncertainty as to the meaning of death has led some persons to record the fact—somewhat in the form of a will—that they would prefer to die rather than to be kept alive when their brain is comatose. But physicians are in an ethical dilemma because, in their Hippocratic oath, they have sworn "... the regimen I adopt shall be for the benefit of my patients according to my ability and judgment, and not for their hurt or for any wrong. I will give no deadly drug, though it be asked of me..." The oath is further complicated by the possibility of lawsuits (either for using or not using life support systems) which question the doctor's ability and judgment (Malecki, 1985).

Accepting the inevitability of death and facing it as a personal event may begin, early in life, when adults stop shielding children from the impact of death. Too often adults refuse to talk to children about it when it occurs in the family, among relatives, friends, classmates, or elderly volunteers at school. It should be talked about as political and educational events are discussed—sporadically, spontaneously, casually. Death and dying are respectable words and need not be disguised as gone to rest, gone away, passing, sleeping, or visiting another land (Kubler-Ross, 1969; 1979). Religious beliefs are regarded as supportive in the acceptance of death. Kastenbaum (1986) suggests that friends and loved ones, especially those with whom communication is easy can help soften the pain and bewilderment of losing a loved one. It is now realized that the phenomenon of attachment, which is emphasized as a vital factor in infancy is salient throughout life in the form of friendship, marriage, and death. Verbal communication during grief is not absolutely essential; another's presence is helpful. The importance of mere presence is indicated by this story.

> A little girl was late getting home from an errand and the mother asked why. "I was helping Judy."
> "What were you doing?"
> "Well, her doll's head was crushed."
> "How could you help fix that?"
> "I was helping her cry."

One of the burdensome factors in death is the easy arousal of guilt in survivors. They tend to recall occasions on which they could have been more kind, helpful, or accepting during life. When such thoughts occur it is well to discuss them. Guilt is less likely to be generated if death has been discussed beforehand—death's inevitability, what happens to property, the body, how surviving family will live, and dispersal of personal effects. Such discussion not only helps the survivors but it tends to make life more highly prized while persons are still alive.

Chapter 5

PURSUING LIFELONG LEARNING

Until recently the need for education was seen as limited primarily to children and adolescents. Schools were expected to provide them with the knowledge and skills they would require for the rest of their lives. Few people today would agree with this formula for success. Instead, there is a growing emphasis on self-improvement at every age level. The emerging attitude is to regard personal growth as a never-ending commitment. This fundamental revision in the public attitude toward learning presents educators with new opportunities and greater responsibility. In this chapter we will examine some: (1) reasons for developing a lifespan perspective of learning; (2) aspects of schooling that encourage continuous growth; (3) ways to improve the learning process; and (4) curriculum needs students have at various ages.

Learning in a Past-Oriented Society

When the older people of today were children, the world was changing less rapidly. Because there was a slower rate of progress, the past dominated the present. Consequently, the young learned mostly from adults. In those days a father might reasonably say to his son,

"Let me tell you about life and what to expect. I'll give you the benefit of my experience. Now, when I was your age..." In this type of society the father's advice would be relevant because he had already confronted most of the situations his son would one day have to face. Because of the slow rate of change, children could see their future as they observed the day-to-day activities of parents and grandparents.

There are still some past-oriented societies in the world, places where adults remain the only important source of a child's education. This continues to be the case among aboriginal tribes in Australia as well as for religious groups like the Amish and Mennonites of America. In each of these settings the future is essentially a repetition of the past. Thus it is justifiable to teach the young that they should adopt the lifestyle of their elders. For this reason—in every slow-changing culture—grandparents are viewed as experts, as authorities, and considered to be models by all age groups. The role expected of children is to be listeners and observers, to be seen but not heard.

Learning in a Present-Oriented Society

Something happens to a society when technology is introduced and begins to accelerate. There is a corresponding increase in the rate of social change. Long-standing lifestyles are permanently modified. Successive generations of grandparents, parents, and children come to have fewer experiences in common. In effect, the children of today are having experiences that were not part of their parents' upbringing. This means there are some things adults are too old to know simply because they are not growing up now. Certain of the situations children presently encounter are unique in history to their age group. Access to drugs, life in a single parent family, computer involvement and global awareness are common among children. Adults cannot remember most of these situations because they never happened to them.

The memory of one's own childhood as a basis for offering advice ("When I was your age...") becomes less credible as the pace of change quickens and young people's experiences begin to diverge significantly from those of their parents and other adults. As a "generation gap" opens up, a peer culture develops. The present period is a time during which adults and children learn mostly from others who are at the same stage of life as themselves. Even a casual observer of America's age-segregated environment will note that peers are turning more and more to one another for advice. Peer influence is strong within every age group. We are just beginning to learn something about the pressure that older adults impose on one another, particularly when they leave their home towns and move to retirement communities.

Obviously a peer orientation undermines cultural continuity as it divides the population into special interest groups. Furthermore, because a technological society assigns greater importance to the present than the past, older people cease to be seen as models for everyone. Instead, each age segment chooses to identify with well known individuals of its own or next higher age group. So, in a technological nation respect for the elderly declines. Older people are no longer regarded as experts about much of anything except aging.

Learning in a Future-Oriented Society

The phase of civilization we are entering is commonly referred to as the information age. In this context schooling for children begins much earlier, continues longer, requires computer literacy and includes considerable knowledge which was unavailable to previous generations. It is estimated that knowledge is currently doubling every eight years. Given these conditions, the young are bound to view the world from a different perspective. As a result, the gap in experience between age groups keeps growing wider, so much so that intergenerational misunderstanding become common.

Another fundamental change accompanies high technology. Life is increasingly dominated by concerns about the future, e.g. preventing nuclear war, preserving the environment and preparing for personal retirement. Certainly it is wise for each of us to look ahead, set goals and make plans—as long as we also attend to current affairs. For those who become preoccupied with the future, there is no longer any spontaneity; everything must be planned. They may be reluctant even to visit friends without giving them ample notice. Life itself becomes just one long period of anticipation. These people are forever looking forward to the time when they graduate, get a job, have children, receive a promotion, save enough to travel. They avoid living in the present. Instead they wait, they save, they are going to do so many things, have such good times someday—and life goes by.

Perhaps you have known parents who live for tomorrow. They look forward to the freedom and economic security that will come after their children grow up. Many of them look ahead to leaving the job, to retirement and leisure time. But in their anticipation of the future, they fail to enjoy their children now, their jobs now, their lives together now—and then when the days of retirement arrive they regret that the children are gone, they resent the loss of their vocational role and they painfully face their failure to have established intimacy. It is just as possible to get stuck in the future as in the past or the present.

At the societal level the high technology emphasis on planning is complicated by our continuation of age-segregation. That is, too many people tend to think about the future primarily in relation to their own peer group. This practice should be revised in favor of a broader personal commitment to the entire society—people of every age, ethnicity, and religion. In a future-oriented society the young and old either learn from each other or exclude some age group from planning the future. It is this intergenerational learning that society must continue to encourage (Lytle, 1985; Thorp, 1985; Tice, 1985).

Schooling in the Past

The kind of schooling each of us encountered was influential in shaping the attitudes we have toward ourselves and others. For example, in a past-oriented society, the one in which those of us who are middle-aged or older grew up in, the main emphasis in classrooms was on analytic thinking. Students considered events that had already happened in the past. They tried to analyze these events, figure out what could have been done differently and, in hindsight, judge how certain situations might have been avoided. The warning was that "Those who do not understand the mistakes of the past are destined to repeat them." Therefore, knowing about the past was considered the best preparation for the future. By studying people and happenings of bygone days, Americans could learn who they were, understand their national identity and what their country stood for. Thus subjects like history and literature (especially the classics) were given special priority. Students were obligated to take American history several times during the compulsory school years to ensure an attitude of patriotism.

Schooling and the Present

More recently, in our present-oriented society, schools have had to take seriously the effects of technology. Alvin Toffler (1970; 1980) told us that we would soon live in a time of overchoice. Perhaps the most familiar example of dealing with overchoice is the way a person feels in a Baskin-Robbins ice cream store with many flavors to choose from; or when confronted by 60 cable television channels. Our children will experience overchoice often so they need to learn to deal with it. They cannot be prepared to do so by limiting their studies to the decisions of persons in the past or having decisions made for them by teachers, parents, and caretakers.

As we enter the era of overchoice, children must have firsthand experiences with critical thinking in order to make good decisions. Although most parents would agree that growing up and decision making go together, it does not always turn out that way. Instead, adults usually justify their takeover of decisions by insisting that children's wrong choices could bring serious consequences. This presents a dilemma. Boys and girls need preparation in making choices but the chance to do so is often denied them for their own protection. Because children are expected to remain students longer than their parents were, the initial chance to decide for themselves may not come until the stakes are high (Chance, 1986).

The importance of making one's own decisions has been reinforced by acceptance of the self concept movement. A guiding premise of this movement is that every person is special, one of a kind. Therefore, each individual should be encouraged to make his or her own choices. As this view grew popular in the late 1970s, schools added critical thinking to the list of goals for learning.

Schooling and the Future

Schools have also been obliged to accept more responsibility for helping students cope with the future. Students can read about problems of the past but they cannot do anything about them. They cannot stop Napoleon from going into Russia. They cannot buy Hitler's paintings so he will become an artist instead of a dictator. They cannot effect these issues but they can influence the future by thinking about it and learning to plan cooperatively. That is, the physical and social technology are available to make the future what we wish it to become. But first there is a need to conceptualize alternative futures and then decide which of these to pursue. On both the personal and societal level, planning for the future requires creative thinking (Torrance, 1986).

Futurists agree that if children are encouraged to be creative, they can more readily avoid boredom, resolve personal conflict, avert war, improve the national quality of life, cope with increasing consumer choice, accept complexity and ambiguity, make independent judgments, use increasing leisure time constructively, adjust to the rapid development of new knowledge and sustain mental health. In an era of overchoice, there is also good reason for education to include moral development, our commitment to others. It is not just that a multitude of options can overwhelm people if their learning lacks a coherent set of values; but as technology makes a greater range of goals possible, we are forced to decide what goals ought to be achieved and the priority of their importance (DeBono 1984; Naisbitt, 1985).

Learning to Reintegrate Society

In order properly to shape the future in a democratic society, intergenerational dialogue has to become more common. Unless such contacts can be sustained and mutually beneficial, the future may bring conflict as low birth rates provide fewer working-age taxpayers to meet the needs of a growing elderly population. Indeed some social scientists expect the opposition between young and old to replace the relation between classes and races as the dominant social issue during the next half century. Whether these forecasts will come true remains to be seen. It is certain, however, that children and adolescents are depending less than ever on adults for interaction.

Fortunately, the potential for better relationships has been recognized. During the past several years there has been local and congressional support for intergenerational programs throughout the country (Stamstead, 1985). These programs acknowledge that it is not enough to just extend life for older people—their lives must be enriched as well. In order to do so, the elderly need something more than a decent pension and competent health care. They also need younger people to spend time with them. This is because at every age we define our importance in terms of amount of attention others give us and the impact we have on human affairs. When the elderly become isolated they lack influence and soon thereafter regard themselves as insignificant. This is why it is a mistake to suppose that leaving older people by themselves and supporting their independence are one and the same. On the other hand, some elderly people do not realize that they have a reciprocal obligation to share time with the young. Instead some of the elderly inadvertently isolate themselves by accepting the proposition that retirement ends their responsibility to society. Every credible view of human development includes maturity as a goal and identifies mature persons as those who share themselves with others.

Intergenerational programs are based on the premise that each age group is the best source of information about its own experience. Moreover, children and adolescents are the groups older people most need to understand and be understood by. Obviously the growth of intergenerational programs implicate elementary and secondary education. This more comprehensive view of learning urges educators to recognize the teaching potential of sources other than themselves. One of the surest ways to disadvantage children of any income or ethnic group is limit them to a single source of learning. Yet this is exactly what happens when teachers choose not to utilize older volunteers as helpers in the classroom and thereby perpetuate the educational conditions of a past-oriented society.

INFANCY

The Infant Intellect

Researchers have been trying for several decades to determine what newborns know. The conclusion is that babies understand more than has been supposed. At the University of Edinburgh, Bower (1982) conducts approximately 1000 experiments with infants each year. In one of these studies an infant boy and girl were filmed making various movements. Their clothing was exchanged in some of the footage and all apparent evidence of gender was deleted from the edited film. Adult viewers had difficulty telling the babies apart but something about their movements enabled infants who saw the film to more accurately distinguish the boy from the girl. Bower contends that babies can tell the gender of other infants whom they observe and prefer to look at those of their own sex. The origin of this competence remains a mystery.

Other investigators have established that babies can be taught almost from the time of birth. Subjects only twelve days old have shown that they already imitate adults who stick their tongue out at them. Moreover, if a pacifier prevented the expression of imitation, infants remember and stick out their tongue as soon as the pacifier was removed. In response to skeptics who refused to believe these findings the experiment was publicly replicated using newborns less than one hour old. The initial results were confirmed.

Another line of inquiry involves how infants organize their learning. Four month old boys and girls were shown a film in which two toys bounced around in different ways, each with a corresponding sound track. Later a single sound track was played and the babies were able to match it with the correct film. This suggests an innate ability to divide events into categories. Again and again it has been observed that babies demonstrate skills and behaviors that seem to have no basis in previous experience (Friedrich, 1985).

Learning through Play

Given the surprising intellectual competence and learning potential of infants, it is natural that parents wonder how they should go about meeting their educational needs. One clue comes from studies highlighting the importance of a close parent-child relationship. Because a baby's primary needs include affection, interaction, and continuous care, it seems wise to emphasize learning activities that fulfill these conditions while simultaneously contributing to cognitive and physical development. Since play is fun for infants and can be for parents as well, specialists in early learning consider play as the most promising medium for instruction. Games involving rattles, mobiles, soft toys, and mirrors have been invented to facilitate basic skills such as eye focus, hand and eye coordination, and the recognition of differences among similar objects. Besides acquiring new information it is expected that participating infants will also develop positive attitudes about new situations and ways to approach them. When boys and girls at play are helped to perceive the world as a predictable, orderly, and consistent place they gain self-confidence and more readily express curiosity (Caplan, 1982; White, 1985).

None of the authors of infant games recommend that they be used as a formal curriculum. Instead parents are encouraged to aim for mutual enjoyment during playtime. When a boy or girl succeeds with one of the games he or she should not be thought of as having passed a test which requires going on to the next more difficult event. Infants enjoy repetition and they should be allowed to practice skills in order to improve them. But what difference does it make if parents take the time to engage their infants in play? Sparling and Lewis (1984) devised 100 games as the basis for child activities in Project Care, a federally funded intervention program. The games ranging in purpose from language acquisition to building self-esteem were played at the day-care center and during weekly home visits by a research team. By age one infants in the experimental group who were exposed to games attained significantly higher intelligence test scores than a control group who did not participate.

Early and Middle Childhood

Creativity and Solitude

The importance of creative thinking in a future-oriented society is generally accepted. But the childrearing conditions for developing this strength should be more widely understood. Over the past two decades numerous attempts have been made to determine what creative people are like and how their extraordinary abilities can become more common. In addition to being highly imaginative, most creative people enjoy solitary activities, can concentrate for extended periods of time, and exhibit an unusual level of task persistence. But how did they get to be this way? What growing up experiences did they have in common? Some clues are provided by autobiographical narratives. Usually, because they were either the eldest child in a family or were distantly spaced from brothers and sisters, they spent more time alone and with adults than did their peers, and they learned at any early age to enjoy the company of their own imaginations (Stewart, 1985).

Although the privacy most creative adults experienced during childhood is suggestive of an environment other people may need in order to become more creative, the value of solitary play was overlooked until Singer (1973) and his colleagues at Yale University made some important discoveries. One experiment involved nine-year-olds who were similar in intelligence, level of education, and social background. After intensive interviews, the children were divided into two groups. The so-called High Daydreamers were boys and girls who created imaginary companions, enjoyed playing by themselves, and reported more daydreams. Children qualified as Low Daydreamers if they preferred more literal play, expressed disinterest in solitary activities, and reported infrequent daydreams. All the youngsters were told that they were being considered as potential astronauts. The candidates were further informed that, because astronauts must spend long periods of time in a space capsule without moving about much or speaking to others, the purpose of the experiment was to see how long they could sit quietly without talking to the experimenter.

The results were significant. The High Daydreamers were able to remain quiet for long periods and to persist in the experiment without giving up—factors closely related to concentration ability. In addition, compared to Low Daydreamers, the High Daydreamers were less restless and less eager to end the test. They seemed serenely able to occupy themselves inwardly to make time pass. Later it was discovered that they each had transformed the situation of forced compliance into a fantasy game which helped them increase their waiting time. In contrast, the Low Daydreamers seemed unable to settle down. They would repeatedly leap up and ask "Is the time up yet?" They continually tried to involve the experimenter in conversation. Further testing revealed that the High Daydreamers were also superior in creativity, storytelling, and the need for achievement.

Learning to Be Alone

Teachers and parents can confirm the solitude findings for themselves. It is a common observation that children from low-income neighborhoods are usually crowded together and seldom have a chance for solitary play. As a result, many of them come to school restless, unable to sit still or persist at tasks, and inclined to act out impulses rather than reflect on them. Because these children tend to lack the ability to concentrate, they often interrupt and distract one another. Many central-city teachers complain that their students have not developed the inner resources needed for sustained study. The situation is not much better in those higher-income homes where children are heard to complain "There's nothing to do" when playmates or television are unavailable. It is a sad commentary when the very young already bore themselves.

Fortunately, if solitary play is permitted a larger place in education during early and middle childhood, boredom and inattention may become far less common. The transition will not be easy because grownups have traditionally felt that children do not need privacy. On the contrary, children are often led to feel that being alone is more a deprivation than it is an opportunity. When parents say "Go to your room and stay there," solitude is represented as a punishment, a type of solitary confinement. Then too, although adults resent intrusions by children when they are trying to concentrate, their tendency to underestimate the seriousness of children playing alone is nearly universal. Yet it has been determined that the frustration effects of interrupted solitary play include reduction of a child's persistence at mental tasks and a lowering of the ability to concentrate. The younger the child, the more vulnerable to play disturbance he or she will be. This finding comes as no surprise to the many day-care and preschool teachers who express disappointment about being unable to grant the request of some children for periodic privacy (Strom, 1981).

Unlike learning to be with age-mates, which is a socialization skill that can be acquired after a child starts school, learning to be alone probably begins at home or not at all. Obviously, the number of children in a classroom and the frequency of interruptions make solitary play a low-priority activity at school. Moreover, once the school years begin many organized groups are available, along with other peer pressures to spend less time alone. Some of these group experiences can offer a child much and should be encouraged. At the same time, however, parents should help children arrange a schedule that allows ample opportunity for solitude each day. This will be especially difficult for men and women who have been unsuccessful in finding uninterrupted time for their own important affairs.

When children engage in solitary play, they fantasize more than during peer play or parent-child play. Still, even if we recognize solitude as the best condition for fantasy practice, we sometimes look with concern upon the child who prefers to play alone. This apprehension relates to the high esteem our culture attaches to extraversion and sociability. Indeed, we generally tend to ignore the evidence suggesting that two-thirds of creative people are introverts. Some parents even express anxious reservation about whether their children's fantasy play is healthy. As one father remarked, "Playing alone is fine, but our four-year-old seems to be a victim of hallucinations. He often makes reference to Roy, a nonexistent companion." If he had listened more closely, this father might have realized that, unlike a victim of hallucination who is controlled by them, during solitary play the child is in control of the imaginary companion.

A better future for all is likely if more parents and teachers come to see the benefits of solitary imaginative play for the development of concentration, task persistence and self-reliance. Then the emphasis on getting to know others will be joined by an enthusiasm on getting to know oneself. Surely children must learn to relate effectively with peers; but, unless they also learn the productive use of privacy, they will have little to offer when in the company of others. To achieve either goal requires that both of them receive support.

Education and Creativity

Probably everyone possesses creativity to some degree. Most of what boys and girls learn prior to attending school comes through guessing, questioning, searching, manipulating and playing. These activities fit most definitions of the creative process. Given the natural creativity of children, educators' first concern ought to be preserving this potential. Creativity will develop if teachers encourage it, but there is a history of support for the claim that many children unnecessarily sacrifice imagination at about the fourth grade. At that point the previously rising creative growth curves begin to decline. This problem was long ignored, because it was thought to be a developmental phenomenon instead of a cultural one. Then it was learned that in certain cultures the development of creative thinking abilities is continuous. Even in the United States no slump occurs at grade 4 under teachers who encourage creative boys and girls (Torrance, 1967; 1979).

Teachers should consider two observations based on a decade of creativity studies at Brandeis University (McCabe, 1985):

1. The single greatest motivator for creativity is freedom. Something happens when teachers accommodate student interests, periodically allowing them to decide what they are going to do and how to do it. This condition of self-control increases the possibility that boys and girls will explore unlikely paths, take risks, and in the end produce something unique and useful.

2. Frequent evaluation and criticism smothers creativity. This conclusion differs from the assumption of teacher trainers who advocate continuous feedback to students. When the desired outcome is creative behavior, people do best if their output is not reviewed frequently. Adults who are the most inhibited, unable to express themselves in imaginative ways, are those in high-pressure situations where they have weekly or monthly evaluations.

Later Childhood

Peer Teaching and Learning

The gap between the experiences of youngsters and adults means that an increasing proportion of what children learn comes from peers. This shift in the source boys and girls prefer to rely on for their guidance has implications for school, especially the way in which instruction is provided. Some observers believe that peer teaching in the classroom is one of the most needed improvements in American education (Cohen, Kulik & Kulik, 1984).

Peer teaching probably began with the one-room schoolhouse. Although the practice of using older and more accomplished students as helpers was not always successful, it did recognize the capability of children to teach peers. What was not realized for a long time was that tutoring can result in important learning for the tutor. Then, in the 1960s President Kennedy called on educators to find new ways to help potential dropouts remain in school. He pointed out that, because of automation, drastic changes were anticipated in the labor market. Many of the unskilled jobs typically filled by dropouts would no longer be available. Something had to be done to reduce the national dropout rate of nearly one million students per year. In response to this appeal, a variety of innovative peer teaching programs were established.

One of these programs was the Mobilization for Youth Project, a massive effort intended to help elementary students in New York City who were having trouble with basic skills. Usually the volunteer tutors were high school students. In order to determine the outcome of these helping relationships, Cloward (1976) and his associates studied changes in achievement test scores. They found that the tutors were making striking gains in test scores, gains much larger than those recorded by the elementary students that they were tutoring. During a five-month period, the boys and girls with reading difficulties who were being tutored gained an average of six months in reading level; the tutors gained a phenomenal 3.4 years.

While the unforeseen benefits of tutoring for the peer teacher were certainly welcome, clearly it was not enough for the tutor alone to make striking gains. Accordingly, Lippitt (1975) suggested that the growing acceptance of peer teaching should be accompanied by carefully designed tutor preparation. Properly guided, tutors could become more effective teachers and receive feedback about the worthwhileness of their efforts. Working with elementary and junior high schools in Detroit, Lippitt designed a cross-age helper program. In this procedure upper grade students are helped to understand what younger children are like and what causes some of them to have difficulty in school. Each tutor learns how to cooperate with the younger child's classroom teacher as well as how to help the child directly. The program spread rapidly and remains popular because the common result for both parties include higher achievement scores and greater self-confidence. In addition, the students being assisted show favorable attitude changes toward learning and receiving help, while the tutors gain valuable experience in understanding others.

The extent to which educators support peer teaching depends in part on how they interpret its effects on their own role. Some who object to the practice do so out of fear that peer teaching will lessen their control and influence. Others admit they are concerned by the prospect of not being the center of student attention. In their opinion, peer teaching could erode the respect that should be reserved for adult teachers. Some argue that children deserve the best instruction possible and that only qualified and trained adults can offer this kind of learning.

Certainly the teacher's position changes as students assume more responsibility for instruction. But this effect can be viewed in another way; every child can use more individual help than any one teacher can normally provide. The key to individualized instruction is more helpers. Further, a common goal of education is to foster the independence of the individual, a goal more likely to be reached when peer teaching is utilized than when everyone in the class must look to a single source for all of their learning.

Parents need to know that peer teaching can result in greater academic achievement and improved self-image for both parties. In addition, more than any other activity at school, peer teaching involves students in responsible tasks on behalf of others. This maturity benefit appeals to an increasing number of parents who can see the dangers of teaching their children to become self-centered. They realize that there are many adults whose training and income qualify them as successful but whose lack of concern for other people labels them as failures. The fact that more employees are fired because of interpersonal relations problems than because of job incompetence underscores the need for schools and the family to facilitate the emotional and interpersonal aspects of growing up. By becoming tutors, boys and girls get to practice valuable human-relations skills and they learn to define success in broader terms.

Improving Homework Practices

Gallup polls during the past decade have repeatedly shown that the general public feels students in elementary and high school are not required to work hard enough. According to teachers there simply isn't enough time during the normal school day to accomplish all of the public mandates for instruction. So, teachers respond in the only way they can, extending the school day by sending work home. Although this practice can be viewed as a way to guarantee more parental involvement and shared accountability, there are other less favorable consequences that deserve attention if the motivation of children for lifelong learning is to be preserved (Pendergrass, 1985).

In many families homework is the source of considerable conflict as parents wonder: Should we pressure our child to do this assignment as a way of showing support for the school and teacher? There are frequent arguments about the worthwhileness of assignments. Many children (and their parents too) feel they have a need to do other things—hobbies, sports, chores, and family activities. And for some students, alienation from school grows because it seems they're never done with homework. Despite these objections homework is not likely to disappear; neither should it be seen only in terms of limitations. When teachers plan wisely, homework can yield benefits for children and parents. The following recommendations are intended to support better assignments by teachers.

(1) The purpose of most homework should be to exercise skills that students have already acquired. To be sure they understand the assignment, teachers should provide an opportunity to ask questions. It is also a good idea to have students do several of the intended homework problems in class before an assignment. This assessment can determine which students are ready for practice and who needs further instruction rather than homework. It is unreasonable to ask students who cannot do problems correctly in class to reinforce the same errors through homework.

(2) Homework that has a distant yet definite due date gives children flexibility in accommodating their out of school commitments. Parents who are aware of deadlines can help by arranging family activities or plans and assisting students (especially young ones) to

schedule time accordingly. When assignments are made for the following day, children and their parents need to know how long a work period is expected by the teacher. Then, at the end of the designated time, e.g. 30 minutes, the homework is put away regardless of whether some problems remain unfinished. This practice permits leisure activities (which are also educative) for some boys and girls who might otherwise have to work beyond their bedtime.

(3) It should be understood that family activities occasionally may require that a child skip homework. To accommodate these inevitable situations each family should be provided five blue slips every semester which, when signed by the parent, entitles the student to an excused assignment. This innovative practice demonstrates school trust in parental discretion and allows mothers and fathers more control of their family schedule. It also helps sustain the respect children have for their parents. A by-product of this strategy is better parent-teacher relationships.

(4) Parents can become more aware of their child's academic progress by sharing the responsibility for checking homework. Picture this situation: Your sixth-grade daughter comes to you close to bedtime requesting a check of 40 mixed-fraction problems. Instead of having to figure out the solutions yourself, you have been provided answer sheets by the teacher. This procedure saves parents time, reduces their embarrassment when they are unable to do the assignments, and offers immediate feedback to students. It also conveys teachers' respect for the family's role in evaluation.

(5) Nowhere is consideration of the individual more important than in homework because the teacher does not control conditions of learning at home. For example, students of every age level are adversely influenced by family dissolution. When this occurs at the college level a student might be advised to reduce the credit hour load or take a semester off since preoccupation with family difficulties could otherwise result in poor grades and put them on academic probation. However, when a fifth-grader's parents decide to split up, the youngster cannot be counseled to take a year off from school. The temporary emotional disturbance of such students is predictable and teachers can help them most by making shorter or less frequent homework assignments with longer time limits.

(6) Homework policies are seldom examined rationally. It is just supposed that whatever teachers decide to assign must be right. Because parents are more aware of what really takes place during

homework they should periodically be invited to evaluate the process and its results. An effective evaluation form provides constructive feedback, helps teachers improve their influence and maintains parent anonymity.

(7) Students in the intermediate grades and beyond should have their homework assignments coordinated so they are not overwhelmed. Unlike primary children who have only one classroom teacher, these students may receive instruction from several faculty members. Often none of the teachers is aware of what the others expect in terms of homework, and students might be given three or four assignments due on the same day. When demands of this kind are continuous, conscientious students are being subjected to unreasonable pressure and deprived of leisure time. Students consider school as their work; they spend six to eight hours a day involved in serious business. Parents and students ought to be consulted in developing the school's homework plan which might include a maximum number of assignments, provisions for flexible deadlines, and a procedure for monitoring the operation.

Learning About Aging

According to the American Medical Association (1983), it is reasonable to suppose that a high percentage of students now in elementary school will live past age 90 and many beyond 100 years. By the time today's fifth-graders retire, in about 2040, the proportion of elderly people in the United States will have doubled and represent 20 percent of the entire population.

The school has always had a role in preparing students for the future but until recently this obligation was limited to developing readiness for an occupation. The way children view aging can determine whether their perspective of the many years that remain are favorable. Gale (1985) reports three curriculum goals for fifth- and sixth-graders in the Chicago Public Schools:

Goal 1: Learning to accept the elderly and get along with them.

Goal 2: To study the process of aging and to look ahead with confidence to the next stage.

Goal 3: To make plans for a long life with an extended period of maturity.

Adolescence

Instrumental and Expressive Curriculum

Teenagers need to achieve a sense of identity and maintain a favorable self-impression. When these goals are supported by teachers and parents, the chances are better for learning to become a lifelong process. Helping adolescents grow can begin by recognizing that the school has two cultures, one expressive and the other instrumental. The instrumental culture is made up of those activities that lead to the skills, knowledge, and values that are commonly stated to be the goals of schooling. A student participates in the instrumental culture for the sake of obtaining or achieving something beyond the mere experience of study and learning—to be promoted, to be praised by the family, to get a diploma, to become a doctor or an engineer. The school consists in part of a set of instrumental procedures that produce an individual who can read, do mathematics, or master a foreign language. But the school also includes an expressive culture, which is made up of those activities that students take part in for the sake of the activity itself rather than for some purpose beyond the activity. Intramural athletics, art, drama, and music are all part of the expressive culture.

Although the expressive culture sometimes involves the learning of knowledge and skills, these outcomes are generally not so important as they are in the instrumental culture. Students can enjoy singing in the school choir, working with clay, helping with a drama performance or playing volleyball without acquiring outstanding knowledge or skill. The criterion of success is not so much how well they do as how much they enjoy doing it. To be sure, the instrumental culture sometimes has expressive impact—for instance, when a student learns to enjoy science experiments, reading novels, or doing algebra problems, and does these things for their own sake. Conversely, there is an instrumental undertone to most expressive activities. But students usually distinguish between the expressive and instrumental activities. Because most students can succeed in the expressive culture, where there is not such rigorous and explicit competition, this realm is especially good for teaching values. In this context, some youngsters can learn to appreciate school, become committed to learning, and to value participation. Similarly, many parents who have little formal education can more easily take part in the expressive aspects of school without embarrassment.

School Success and Failure

There are various combinations of success and failure in the expressive and instrumental cultures of the school. The social and intellectual development of a student depends of his particular combination of successes. A high level of success in both cultures is characteristic of the good student who also finds satisfaction as a social leader. A high instrumental/low expressive combination describes the bookworm who is socially invisible but does well at scholarly pursuits. A low instrumental/high expressive combination defines the mediocre student who is able to obtain satisfaction at school because of participation in extracurricular activities. (There are young people whose interest in athletics or band is the only incentive for staying in school.) Finally, a low instrumental/low expressive combination represents the student who can find little satisfaction or reward in the school. Often such youngsters quit formal education as soon as possible. The dropout problem is most acute among minority groups. Blacks quit school at nearly twice the rate of whites and Hispanics leave at almost three times the rate of whites (Carnegie, 1985).

Creative use of the expressive culture can make the school a place of satisfaction for more students as well as improve performance within the instrumental culture. A dramatic illustration of these possibilities was recorded in Junior High School 43, located in a slum section of Harlem. About 50 percent of the school's students were black, and nearly 40 percent were Puerto Rican. The most successful half from the entire school population of 1400 were selected to participate in the Higher Horizons Project (Hillson & Myers, 1963). These students had a median IQ of 95 and were, on the average, a year and half below grade level in reading and mathematics. Instead of increasing the amount of time these students spent on math or reading by eliminating the so-called "frill" areas of education, the project took the opposite approach. That is, a number of motivating influences from the expressive culture were introduced to increase satisfaction with school and desire for learning. The boys and girls were brought to sporting events, concerts, and movies. They went to parks and museums and on sightseeing trips.

Follow-up studies showed that the project members who entered senior high school graduated in substantially greater numbers than had those from the same junior high school in pre-project years. Furthermore, 168 of the graduates went on to institutions of higher education, compared with 47 from the three classes preceding the experiment. It seems that the expressive activities gave rise to some important value changes in the experimental group that in turn led to improved performance in the instrumental culture (Strom, 1975). Findings like these ought to cause more of us to take the extracurricular, the nonacademic, the expressive aspects of schooling far more seriously than we do.

A Broader View of Basic Curriculum

It seems clear that many parents who are anxious about the occupational futures of their children do not recognize the need for a balance of academic and expressive activities. Consequently, they often urge the school to reduce its expressive offerings in favor of a return to "the basics." In turn, when teachers are held accountable for scholastic achievement only, other important aspects of a student's personal development are neglected. For example, although it is seldom mentioned in this context, racial integration is one of the basics declared by our courts. Unfortunately, the "back to basics" philosophy of deemphasizing expressive subjects reduces the possibility for integration. Different ethnic groups do not come together just by being bused to the same school; nor is the instrumental curriculum expected to promote social interaction. This leaves the expressive culture as the realm in which young people best learn to appreciate and value others. Here they engage in mutually satisfying, cooperative activities that permit spontaneous and sustained communication, whether through a new course on ways to use leisure or through the more traditional courses in physical education, music, and art. When this kind of involvement enables a teenager to begin a friendship with someone of another race, it can do more to offset racial prejudice than any other school experience. By encouraging participation in the expressive culture, parents and teachers can help students develop self-esteem, identity, and mental health.

EARLY AND MIDDLE ADULTHOOD

Higher Education in Adulthood

Colleges have traditionally catered to students between 18 and 22 years of age. But, owing to a low birth rate, the projected enrollment of this segment is expected to drop 20 percent by the year 2000. During the interim, persons in the early adult years will become the nation's largest subpopulation. The shift to a larger proportion of adults in higher education has already begun. Some of the factors motivating their return to school are greater longevity, influence of the women's movement and a rise in job qualifications. In order to accommodate the growing number of adult students many institutions have begun to modify advisement and counseling practices, scheduling procedures, curriculum content, and instructional methods. The results will not be known for some time but there is evidence that progress is taking place (Allcorn, 1985).

More than two-thirds of the college reentry students are women. Often these persons are homemakers who have not attented classes for 20 years. They are excited about returning to school but anxious and sometimes lacking in self-confidence. Depending upon the individual, the initial counseling needs many include an orientation to career retraining, arranging for tutorial help, learning how to develop study habits, use the library, assistance with stress management, and advice about handling multiple roles or issues of life transition. Because these problems differ from those faced by traditional undergraduates, it is essential that the staff who offer counseling services are equipped to help the reentry population.

Scheduling classes may not seem difficult to young full-time students who live in a dormitory on campus. But for persons with a job or responsibility for childcare, matters of timing may determine whether it is possible to even attend college. Alternative scheduling patterns are emerging. The Wednesday college concept appeals to homemakers because a single day of continuous classes each week allows them to make family arrangements. Some adult degree programs offer courses year round in eight-week intensive sessions. Independent study courses, off-campus classes and night school scheduling are especially appreciated by students who are also employed fulltime.

The differing motivation of adult students requires that new courses be developed for them. Although they are expected to enroll in the existing curriculum as well, much can be gained from short-term seminars on topics such as: Introduction to Liberal Arts; Understanding Adult Development; and Personal Learning style assessment. Given the range of competence among returning adult students, it is appropriate for the faculty to consider proficiency testing and other means of assigning credit for prior learning (Wolf, 1985).

Universities and colleges vary in the extent to which they are willing to make changes for adult students. Some faculty members may resent having to modify their usual daytime class load to include meeting students at a less convenient time or location. Others dislike the idea of developing new courses which they consider to be less rigorous than traditional offerings. The erosion of academic standards is cited by some professors as their reason for opposing any deviation from the classical curriculum, teaching practices or methods of assessing learning. These sorts of resistance are predictable and can best be overcome by providing inservice training to the faculty. They should become aware of the enlarging mission of higher education, the needs of adult students and ways to successfully serve them (Peterson, 1983).

Self-concept and attitudes toward learning are as significant in adult learning as during adolescence. Adults can choose not to be in classes where they are subject to embarrassment, humiliation, sarcasm, or discourtesy. Further, teachers of adults need to repair the damage done to self-concepts by competitive grading systems. They must concede that adults come to learn and do not need to be motivated by the threat of low grades or failure. In order to encourage these learners, their special competencies and experiences can be turned to advantage by the professor. Although adolescents are a captive audience (by virtue of compulsory attendance laws) the same principles of self-worth, teacher courtesy, motivation, and varied learning styles determine whether a student pays attention or just attends class. Teacher attitudes are also a factor in adult learning. Adult students will benefit if teachers believe in them. Adults who choose to return to school, who admit a need for help, and who choose to change a lifestyle or become more job proficient deserve their teacher's respect.

THE CORPORATE CLASSROOM

The Carnegie Foundation for the Advancement of Teaching describes a growing commitment by U.S. corporations to provide education for the workplace (Eurich, 1985). Given the rapid change in knowledge, business leaders have determined that even recent college graduates are soon out of touch. Therefore, in order to survive in the competitive market, corporations are educating 8 million employees, almost as many people as are enrolled in colleges and universities. It is not surprising that the corporate classroom is regarded as a significant force in adult education. Some subjects like financial management resemble those offered in traditional schools while other courses such as advanced nuclear engineering compensate for gaps in the standard college curriculum. It is generally agreed within the business community that university courses lag from one to three years behind events taking place in the job world.

Corporations spend $40 billion a year (two thirds of the combined budget for all colleges and universities) to ensure that their employees have the necessary knowledge and skills. Many major companies like AT&T, IBM, and Xerox maintain a national headquarters solely for purposes of coordinating educational programs. The Carnegie study suggests that the quality of industry-based education is often superior to what a student might obtain at a university. Certainly the goals of instruction are more clearly tied to application on the job.

A growing number of businesses have attained an academic legitimacy for their educational programs, often subjecting them to the same criteria for standards and accreditation review that obtains for academic institutions. Regardless of where a course is held, the employees get most or all of their instruction free and often get paid while they attend. This arrangement is so essential for company productivity that some high technology firms consider it appropriate to assign between 15 to 20 percent of an employee's paid time to continuous learning.

A new era of collaboration between businesses and universities seems inevitable. For example, the National Technical University is the most innovative of the degree granting corporate classrooms. It organizes the videotaping of advanced engineering classes at 16 co-operating universities. When the tapes are received at the business site, the working engineers are scheduled to attend the lectures via VCR and then mail their course work to the originating school. The credits taken from the various schools are assembled by National Technical University which awards a masters degree. In some cases satellite transmission and teleconferencing are also included so that students can participate in dialogue. Certainly there is a difference between education for profit by a corporation and the scope of learning universities are expected to provide. Nevertheless, the Carnegie study urges academic institutions to undertake a careful self-appraisal based on the corporate classroom.

Readiness for Conflict

During adulthood most people find it necessary to cope with new types of conflict situations. Some of these disputes arise from adjusting to life with a spouse, trying to bring up children, relationships with coworkers and neighbors. Although disagreement and confrontation have been experienced since childhood, there are some important new factors which must be taken into account. First, unlike the past when interpersonal differences were mediated by family members or school officials, full responsibility now rests with the individual. And, for parents there is the responsibility of teaching their children how to conduct conflict. When boys and girls do not learn from their family how to settle arguments, they tend to rely on immature peers. This leaves youngsters less prepared for the demands that a conflict culture will make of them. Understanding how to accept and express differences is absolutely essential to the growing up process in a diverse society and its learning cannot be left to chance.

How well prepared are adults to face their conflicts? To what extent are they ready to instruct their children in the skills of resolving disputes? By juxtaposing the traditional and emerging views of conflict it becomes clear that this human relations aspect of learning qualifies as one of the dominant educational needs.

Changing Views of Conflict

Until recently, the general tendency in most cultures has been to take a dim view of conflict. Faith systems, utopias, government, and social sciences all present this impression. Most religions regard salvation as a state of rest in which there are no unsatiated needs, no basic grudges, grievances, or disagreements. Similarly, the visions of earthly paradise that we call utopias usually start from the premise that everyone is essentially alike. According to this simplistic model of human experience, social structures can be designed to ensure universal satisfaction. Although we know that people tend to be unique, changeable, and inconsistent, those who construct utopias are seldom influenced by the diversity of human beings. In all the social sciences, from the emphasis on what happens inside a person to studies dealing with the human condition at the international level, the basic views have often been that people try to escape from conflict and that conflict is necessarily bad.

However, both the social conditions today and the forecasts for tomorrow urge us to look at conflict differently. The fact is we will experience more conflict in the years to come instead of less. In the first place, there is a worldwide breakdown of homogeneity. Previously the idea has been that people living in the same territory should be of the same kind—ethnically alike and, in more modern terms, members of the same nation. But the tendencies toward mixture are increasing much faster that ever before. A rising number of people identify with more than one country because perhaps they have a foreign-born spouse, have studied abroad, have lived overseas, or have worked for an organization like IBM with offices throughout the world. Then there are the international groups and ideological associations that bridge national borders. These dislocations of classical patterns of loyalty are growing rapidly, and they make the way of structuring large-scale conflict less obvious. To make matters even less clear, we are witnessing a global mass migration toward centers of urban development. The net result is that people on almost every continent will increasingly find that their neighbors look different from themselves and act differently as well.

The second reason we can expect increasing conflict stems from the rejection of coercive management. All over the world, institutional dictatorship is being challenged as underdogs unite. Organized groups bring the power of their cohesion to bear against the traditional power of authorities through direct confrontation—in prison, by inmates; at universities, by students; in factories, by labor unions. In a related movement, the self-concept phenomenon which originated in the 1970s caused people of every age and background to view themselves and their opinions as important. This departure from the custom of assigning respect exclusively on the basis of social status increases the likelihood of conflict. Within the family the erosion of adult authority toward a more equality oriented role structure has left many parents ambivalent. They want to encourage assertion by their children and at the same time maintain the final authority for decision making without seeming to be dictatorial.

Another reason we will experience more conflict lies in our relationship to the future. For both nations and individuals, there is great uncertainty and also a sense of new possibilities. Because of technology, it appears to be more in our power to influence the future than ever before. However, as technology makes a wide range of goals possible, we are forced to decide what goals to pursue and the order of their importance. People of differing age groups and political persuasion disagree about the goals we should adopt, the changes that should be made, and the priorities that should be established. Particularly in democratic countries, where diverse opinion is acceptable, there is bound to be continuous conflict (Gaylin, 1984).

In addition to the world breakdown of homogeneity, the erosion of institutional power, and our new relationship to the future, longstanding ways of resolving conflict are breaking down. One customary method of dealing with conflict has been to repress it, partly by denying that it existed, partly by not discovering it. This expedient seems less possible in an open society, where many social scientists are allowed to make the society transparent; if we do not see a conflict, they will describe it to us. Wars have moved toward the prospect of total annihilation, but we have made too little progress in preparing ourselves for nonviolent alternatives.

There is also reason to be optimistic about meeting the challenges arising from increased conflict. A science of conflictology has been emerging that draws on the insights and techniques developed in the social sciences from psychology to international studies. Adults can access much of this new learning in a variety of contexts. For those who choose to attend a college or university, there are many innovative courses available which explore topics such as "Interpersonal Conflict," "Peace Studies," "Domestic Violence," and "Building Successful Relationships." Most of these opportunities include a chance to learn and practice constructive ways of conducting conflict (Fogg, 1985).

Because technology will allow us to shape some aspects of the future, there may be much conflict over which plans are most desirable. To bring about the best tomorrow, courses like "The 21st Century" involve training in the design of alternative futures. Students in these classes are urged to go beyond the problem-solving ethic of looking for fault, limitation, and weakness. They are encouraged to thing about the possibilities for our future, to consider many lifestyles and environmental options. Simulation is used most by colleges of business and engineering to aid students in learning how to solve conflicts without having to suffer the consequences of real-life situation. This same approach is becoming popular in industry for purposes of management training, recruitment, and promotion (DeBono, 1984).

Parents can learn more about resolving domestic conflict through informal classes offered by community center, public schools, and churches. For example, Parent Effectiveness Training is a curriculum of attitudes and skills for solving family conflicts and maintaining good parent-child communication. These short-term workshops, offered throughout the country, are usually sponsored by parent teacher associations. "Assertiveness training" is another popular class, especially among women whose upbringing did not include the development of strategies to support personal independence. Marriage Encounter sessions, usually initiated by churches, can help husbands and wives better understand the needs of their spouse and discover new ways to resolve disagreements. These sessions are intensive, often requiring a full weekend. During the time away from home, parents need not be concerned about their children since they are cared for by another couple representing the marriage encounter organization. Finally, there are an increasing number of self-help groups for persons whose chronic problems with alcohol, drugs, gambling, or family abuse create conflicts which they cannot overcome alone. In combination, these sources of support offer encouragement, skill, and understanding for adults who want to deal more effectively with conflict.

Given the lack of training in conflict management most parents received in school and at home, they are naturally reluctant to assume the role of models for conflict. By their own admission, they just don't know how to work out their own differences very well and are therefore unable to teach what children need to learn. That many adults require greater learning in how to live with differences is apparent. One has only to note the size of domestic court dockets or the incidence of litigation between neighbors to confirm that withdrawal or attack is still the dominant reaction to disagreement. Increasingly people are turning to the single subpopulation that has been schooled in conducting conflict. Rather than learn to resolve one's own conflicts, it is easier to announce "My lawyer will speak for me."

LATER ADULTHOOD

Access to Learning

Educational programs for senior citizens are an important component of the lifelong learning process. While most of society's efforts for this age group have centered on basic or maintenance needs such as housing, healthcare, income, and transportation, there is a growing realization that intellectual stimulation and creative development are also essential for mental health. Much of the educational curriculum for older people, whether provided by colleges, government agencies, senior centers, churches, or long-term care facilities has developed in a fragmentary and haphazard way. There is virtually no coordination among the groups offering services and, unlike education for younger people, no organization is responsible for developing an overall age-appropriate curriculum. Because the capacity of older people to learn has been scientifically established, it is appropriate to move ahead in making lifelong learning available.

Access to learning has been limited by several factors. First, there have been institutional barriers. Until a few years ago, entry into college was reserved for younger persons who scored well on admission tests or who ranked high in the upper half of their graduating class in high school. These criteria kept many senior citizens from applying to college because over half of them did not finish high school. Fortunately this obstacle was removed by most institutions during the 1970s. There have been other barriers as well. Even when higher education began to enroll minorities the initial condition was that earning a degree represented the major motive for learning. Many who chose not to pursue a degree were regarded as lacking the serious intent and willingness to plan needed for success in college. As a result, many of the elderly were obliged to seek learning at other sites than universities (Leich & Pieper, 1982).

Several related movements were helpful. In the mid 1970s Walden, Nova and a number of other so-called "universities without walls" moved into the continuing adult education market. Some of these programs exploited students by promising credit for "life experience" while offering quick but often worthless degrees. The contribution of such institutions was that they offered an alternative to traditional procedures and thereby caused universities to reconsider new ways of serving the adult population. Then, following the introduction of policies to facilitate learning for the handicapped in 1975 and reentry students (women's studies in 1977), the Elderhostel movement was created in 1979. This private, nonprofit group, inspired by the youth hostel and folk schools of Europe is for older citizens on the move—not just in terms of travel but in the sense of reaching out to new experiences. It is based on the belief that retirement does not mean withdrawal, that the later years are an opportunity to learn. There are Elderhostel programs at several hundred universities worldwide. The elder participants live in a dormitory on campus for one or two weeks, usually in the summertime, during which they listen to lectures, engage in discussions and take fieldtrips. The programs which typically focus on social science topics are conducted by regular faculty members. There is no homework, no tests, and no grades (Elderhostel, 1986).

With the exception of Elderhostel programs, most universities do not regard teaching older people as part of the university mission. There are several reasons for this circumstance. First, many state legislatures have enacted a policy of tuition waivers for older people attending higher education. Because of the way resources are allocated in universities the lack of revenue for teaching this population makes them a low priority audience. The reluctance of faculty to excuse older students from the same assignments and testing procedures as their classmates is another factor. There are also significant logistical issues involving transportation, parking, and class scheduling (Peterson, 1983).

Basic Curriculum For the Elderly

Some authors suggest that it really dosen't matter where senior citizens attend classes or whether their teachers have advanced degrees or not. The important thing is that they study what pleases them and they receive mental stimulation (Euster, 1982). We agree that classes can take place at a senior citizens center, nursing home, or any location in the community. Moreover, mental stimulation is better than the absence of learning. But until a curriculum is devised to fit the age-stage learning needs of older people, they will not be well served.

It seems reasonable to begin developing a basic curriculum to make available and recommend for older adults. Some of one's education should emphasize obligations and roles. Consider the 65-year-old woman who wants to improve her family influence as a grandparent. In the present scheme she is advised to take general psychology or personality development on the hunch she can somehow extract from these classes something related to her role. A better approach is to design courses which focus directly on topics of relevance to older people. For the 50 million American grandparents this means discussion classes dealing with issues like:

The Influence of Modern Grandparents; Improving Communication With Grandchildren; The Schooling of Grandchildren; Family Conflict Resolution, Grandparents and Sex Roles, and Grandparents and Children of Divorce. By balancing what they want to learn with what they need to know, older people can continue to remain influential in our society (Strom & Strom, 1985).

Chapter 6

FACILITATING POSITIVE MOTIVATION

Motivation is concerned with the reasons for behavior. It relates to why action is initiated, why it is sustained, and why it is terminated. There are of course many causes, often conflicting, sources of motivational behavior. The study of motivation may answer questions relating to why some people behave in a socially oriented, constructive manner while others direct their energies into self or social destruction.

The concern of this chapter is to (1) learn what motivates people to choose and pursue goals, and (2) understand why some individuals seemingly avoid goals or lack aspirations. Typically, goals are determined by basic human needs (hunger, thirst, companionship) or by society or culture. Currently most people strive to actualize personal potential, become responsible for supporting oneself and children, define and pursue a "better way of life." Some persons seem not to care enough about self or society to pursue the roles of independence and responsibility. Those who pursue self-defined and socially defined goals are said to be motivated. Those who just drift with the currents are said to be unmotivated. To the extent that we understand our own motivation and that of others, some power may be generated to choose goals wisely and be in charge of the direction and speed of personal development.

MEANINGS OF MOTIVATION

Multiple Explanations of Motivation

Motivation is internally and externally generated. Early in the century psychologists attributed motivation to instinct. There was an instinct for about everything one could do—eat, sleep, fight, love, work, learn. The list at one time reached 500 and was crushed by its own weight. The instinct concept explained so much that it really explained nothing. The concept was discarded. But its basic, inborn, congenital nature refused to die. Today what once was instinct is more carefully defined and delimited as biological motivation (Whalen & Simon, 1984). Genetic structure does have some impact on aggressiveness, perceptual ability (visual and cognitive) and predispositions or talents. The neuroendocrine system (hormones, neurotransmitters) influences the strength and persistence of motivation. But biological factors do not alone explain goal-seeking behavior.

The behavioral view is that motivation stems from environmental circumstances and particularly on what is reinforced by experience (Skinner, 1971). For example, children are motivated to learn to read, and to continue reading by parents who read to them, who are themselves readers, and who provide books and other print materials. Some persons dislike the behavioral view because it appears to make developing individuals the pawn of circumstances.

Proactive and cognitive psychologists say that persons can decide to have a choice. Our mothers reflected this view when they said, "You can do it if you just make up your mind to do it." Mothers were not entirely correct. Most of us, no matter if we so chose and practiced diligently will not be able to high jump six feet or run 100 meters in 10 seconds. But choice in motivation cannot be summarily discounted. It seems probable that the mother of Frank and Orville Wright said, "You might just as well turn to something practical. You'll never be able to fly."

Hierarchy of Needs

Maslow (1970) devised a hierarchy of human needs that has been applied to education, psychotherapy, and childrearing. His ideas grew out of discontent with the motivational merit of studying Freudian personality disorders. He felt that a better approach to development would be to focus attention on people who were models of mental health. By observing those persons he termed self-actualizers, Maslow hoped to identify the characteristics that give rise to constructive motivation and maturation. In time this concept of wellness would enlarge and form the basis for what is now called wholistic health.

According to Maslow, we are creatures of ever expanding wants. Once our primary needs are satisfied, others take their place, starting with basic biological requirements and proceeding through a series of levels, each more intangible than the preceding one. As a particular need is met, its urgency declines and the next higher one assumes priority. Thus, the satisfaction of lower order needs becomes the foundation for devoting more time and effort to higher order needs.

Several levels of needs are identified in the hierarchy. The most prominent of these are: (1) physiological needs—food, water, air, shelter, rest, and exercise; (2) safety needs—freedom from fear of deprivation, danger and threat, need for law and order; (3) social needs—to associated with others, to belong and be accepted, to be loved; (4) ego needs—for reputation, status, self-respect and self-esteem and (5) self-actualization needs—the realization of individual potential, liberation of creative talents, the widest possible use of abilities and aptitudes; in short, personal fulfillment.

Maslow considered physiological, safety, social and ego demands as comprising our deficiency needs. Until children have enough to eat, judge their environment to be safe, and feel accepted by others and good about themselves, teachers cannot effectively appeal to their higher-level needs for learning and personal aspiration. At the time of his death, Maslow (1970) estimated that in American society the average citizen satisfied about 85 percent of physiological needs, 70 percent of safety need, 50 percent of social needs, 40 percent of ego needs, but only 10 percent of self-actualization needs. If this record could be improved then more of us, regardless of our academic achievement, could move toward the ideal of self-actualization. Self-actualization means giving up the desire to be better than others in favor of striving to be the best one can be. Because self-actualizers are concerned about the welfare of others, they would be motivated to reduce the societal incidence of unfulfilled deficiency needs.

INFANCY

Biology of Motivation

Since the time psychology became a discrete study, the biochemistry and neurophysiology of behavior, including motivation, has been recognized. Surgical, pharmacological, and electrolytic studies show that there are brain areas that are specifically involved in motivational drives (hunger, thirst, activity). One of the outcomes of such studies is the postulation of the hypothalamic hypothesis. This refers to the idea that they hypothalamus serves a metering function in hunger, thirst, homeostasis, and for many emotions and motives. There is some evidence that malfunction of the hypothalamus may contribute to overeating and obesity. But, the organism acts as a whole and to place too much emphasis on one organ (or brain area) is oversimplifying. Achenback & Edelbrock (1984), for example, emphasize the need for taking biological factors into consideration when studying such phenomena as infant autism and hyperactivity. Food sensitivies, brain damage, and neurotransmitter abnormalities are factors in infant and child behavior.

Caretakers need to admit that the infant has physiological needs, and, especially in illness, chemical imbalances. Chemistry, however, does not do it all. The manner and quality of caretaking also influences the shaping and manifestation of motivation.

Attachment

Attachment—the affective bond babies show to persons (usually mothers) with whom they have a lasting, stable, relationship—is a vital ingredient of infant motivation. Longitudinal studies of motivation show children assessed as secure in their attachment as infants were much more enthusiastic, persistent and expressed more positive feelings than did children who had earlier been assessed as anxiously attached (Sroufe, 1982). The manifestations of secure attachment included motivation to explore in strange situations, enthusiastic eagerness, persistence, and flexibility in performing laboratory tasks, not being readily upset when the caretaker leaves the room, and readily engages in play or talk with a stranger. The manifestations of anxious attachment are (recall we are here concerned with motivation) tending to cling to the caretaker when in new situations or in the presence of strangers, crying or temper tantrums when the caretaker leaves the room, and withdrawal from contact with others.

Because of the helplessness of infants and their total dependence on caregivers, the origins of attachment must be attributed to the caregiver. Experiments with monkeys show that attachment is fostered by contact with warmth and softness. The concept of the security blanket is more than just a cartoonist's fancy. Being fed, kept dry and safe by a loving person who is unhurried and makes eye contact generates attachment. Supportive mothers foster attachment; they interact vocally with the infant, they are consistently dependable, and they respond readily to indications that the baby wants their attention (Brewer & Brewer, 1985; White, 1985).

Anxious attachment is likely to arise from mothers who are worried and preoccupied with other concerns; outside work, discord with husband, financial concerns, or self-doubt. Such mothers may treat their infant in a perfunctory manner, hurrying to get the baby fed, bathed, changed, and put to bed. Unrealistic expectations can also contribute to maternal discontent. These mothers tend to be overhelpful in play or task performance. They tend to be exacting when the child is old enough to work puzzles, assemble toys, and play with other children (Brazelton, 1983).

Reinforcement

Babies are said to be egocentric but, in fact, there is much in their behavior that reveals social consciousness; ceasing to fret when someone comes close, watching a person's face when the opportunity exists, and eagerness to be taken into the arms of anyone available—including the precariousness of being carried by a brother or sister who is only slightly bigger.

Because of this social motivation, normal babies who are active, exploratory, and curious are responsive to reinforcement. When they get attention for a given act, that behavior tends to be repeated. Reinforcement may occur through the gratification of success or through praise, or a token given by another. When the reinforcement is given by another, the reinforcer must get on the level of perception of the person being reinforced. Mothers often wonder why their young children so often misbehave in the presence of company. An insightful observer would be able to explain that this happens because the mother notices, and reprimands the child when others are present. In other situations she tends to ignore the child. From the child's point of view it is better to be punished than to be ignored (Trotter, 1983).

Galloway (1976) has compiled some simple rules of reinforcement: Watch the child to assess correctly the effect of one's reinforcement strategies. Reinforce appropriate responses immediately. Do not overdo—keep the learner "hungry." After initial success, reinforce only intermittently. Use tangible rewards—food, a smile or pat. Reinforce only appropriate responses. Be patient with gradual improvement.

There is a hazard in reinforcement. The learner might become dependent on rewards or praise and somehow fail to gain autonomy. Hence the reinforcer should bear in mind that the ultimate goal is child behavior that is spontaneous and socially oriented. one antidote for the overuse of rewards or praise is to accept and love the child not because he is good but because he is valued as a person.

EARLY CHILDHOOD

Fantasy Play and Values

There are many playthings parents feel children could do without. Some dislike military toys because they reflect violence. Others oppose stunt cycle toys because they encourage risk-taking on bicycles. Similarly, crash cars are thought to sanction a disregard for safety, and of course oriental dolls present martial arts as a method of resolving conflicts. Parents with these complaints are ambivalent in that they want to purchase toys which reflect their own values, but they also recognize the need for children to make some decisions in order to develop a value system. And where is it more appropriate for children to choose than in the realm of play? Grownups can justify making certain decisions in children's behalf, such as whether they are to attend school, whether they will see a doctor, and when it is time for bed. On the other hand, to claim that boys and girls need coherent values and then to deny them choices is unreasonable. So mothers and fathers are naturally bothered about the priority they should give to the feelings of children in the selection of toys.

When parents confront us with this issue, we urge them to reexamine their role. Think of it this way: Instead of declaring your values by choosing the child's toys or by censoring the content of play, enact your values during parent-child play. The imposition of values has less influence than the illustration of values. So, if you don't like war, you can try to include the peacemaker role as part of the play with toy soldiers. Show your values to your child instead of protecting yourself from having to demonstrate them. It isn't a question of who is opposed to war; almost everyone is. Rather, the issue is how to develop a concern for peace.

All of us share the aspiration that either a balance of nuclear power or international disarmament will rid us of war. But, although peace means the end of war, it does not mean the end of differences. Because there is a critical distinction between the fantasy wars of children and the bloody wars of men, it is a serious error to misinterpret the motives of the preschool soldier. The readiness to credit children with criminal intention denies them respect; it also illustrates the need for adults to see favorable possibilities in a child's choice of playthings.

Certain needs of children may be served by conflict toys and games. This kind of play offers relief from feelings of powerlessness and dependence, which together account for much of a child's experience. Surely there is nothing strange about the desire to control others, especially those who daily exercise power over yourself. Children delight when they can assert themselves and make daddy run away or fall down because he's been shot. Grownups often forget that boys and girls realize themselves through the impact they make on the world. Because many parents recognize as worthwhile only those strengths that emerge during adolescence, it follows that little children must experience power vicariously if they are to experience it at all. Then, too, conflict playthings offer a safe setting in which to express disapproved feelings, such as anger, fear, frustration, and jealousy. These are normal feelings that should not be repressed, but in many homes they are met by punishment, ridicule, or shame. Danger-type play also provides an opportunity to repeatedly confront fearful issues, such as war, death, and injury. Although these subjects are of universal concern to young children, many adults try to avoid discussing them and in the process increase a child's anxiety.

Risk-taking requires practice in a low-cost setting. During danger play children can afford to take chances, to see what it is like to rebel, to be the bad guy or outcast—risks they dare not try in their family life. In this connection it is worth noting that for some children war play is the only context in which they can conduct conflict without guilt. Even though parents should teach constructive methods for settling differences, some boys and girls are taught instead to feel guilt whenever they oppose an authority figure. For many kids competition needs are fostered by fighting off the mutual enemy. War play also allows children to experience leadership, to lead and command others as well as to become heroes like their television favorites. Finally, conflict toys and games are enjoyed by the participants—they're fun, a fact that should be appreciated by a pleasure-oriented society.

Instead of placing all our emphasis upon the choice of what we allow children to play with, we can recognize that how we play with them also has an influence. Otherwise, we overestimate the plaything and underestimate the influence of the player. We cannot fulfill our childrearing role merely by buying the right toy or disallowing the wrong toy. Our best role is not that of a judge but of a play partner. It is disappointing that people who believe certain toys can actually have a lifelong disabling effect on the personality of their children often find it difficult to believe that they themselves can have any influence by playing with a child. Yet the latter possibility is supported by research evidence, while the former is not (Strom, 1981).

Parents should be cautious in judging the content of play. Once we insist that the direction of children's play is our choice and not their own, the young are no longer decision makers. And, in fantasy play, choice making is essential for participation. We can share in the direction of play only if we substitute the role of partner for that of judge. It is unwise to interpret the content of children's play as a direct correlate of serious motive. When an adult actor portrays a killer, we may say the performance was convincing and therefore successful. However, when a pretending child chooses to play this same role, the supposed reasons for becoming that particular character may receive more attention than the performance. It is this cynical way of interpreting the content of children's play that leads to unfair inferences and to the attribution of motives that children do not have.

MIDDLE CHILDHOOD

The Pygmalion Effect

Those who work with children should respect and admire them—and tell them so. The Pygmalion Effect refers to the fact that such admiration can influence success. It confirms that persons tend to behave and become what others expect of them. The term comes from a play written by George Bernard Shaw, *Pygmalion*. The play later became the musical *My Fair Lady* in which an English gentleman wagered that he could transform an impolite, unschooled street person into a lady who could be accepted in the upper crust of London society. He won his bet but fell in love with the beautiful person the girl had become.

Rosenthal and Jacobson (1968) postulated that high expectations on the part of teachers would cause pupils to make better use of their potential for achievement. Their success was so startling that some people in the educational community thought that they were exaggerating or falsifying their data. Others sought to verify or deny the hypothesis with their own replicating, or slightly varied experiments. The gist of the Rosenthal and Jacobson study was simply to raise teachers' expectations by telling them that they, the researchers, had unique tests which would identify elementary pupils who were "about to blossom." In fact, they had no such tests—the "about to blossom" pupils were selected at random. Nevertheless at the end of the school year the identified pupils improved their mental and achievement scores to a markedly higher extent than the control group (classmates with similar age, sex, and ability but not selected as blossomers). There were no special classes, no special projects, no special challenges for the blossomers. Rosenthal and Jacobson were sure that the only difference was teachers' expectation.

There are implications in the Pygmalion studies for all pupils—blossomers or not. Teachers must face the possibility, despite their denial, that they do not really treat all pupils alike, that they do expect certain of their pupils to behave, or misbehave, in certain ways. Their talk in teachers' lounges reveals the fact that most have differential expectations—and pupils tend to confirm the expectations. "Extensive research shows that teachers' interaction with students perceived as low achievers is less motivating and less supportive than interaction with students perceived as high achievers" (Kerman, 1979, p. 716). Over a three-year period of inservice teacher workshops, Kerman demonstrated that this almost automatic attitude of preferential treatment of able students could be changed to one of equal opportunity for all to learn.

Motivation in School

The need for teacher respect of students is clear. Yates, Saunders, and Watkins (1980) found that meeting the basic needs of students is a viable foundation for helping those with behavior problems to adjust and succeed in school. In our culture, where there is seemingly unlimited diversity in tasks to be performed, success can be achieved in a great a variety of ways. So too should success for children in school be possible in a variety of ways. Unfortunately, children's success is narrowly judged in terms of how rapidly and well they can learn to read, write, and calculate. In contrast to a wider description and distribution of success there is presently a great cry for "getting back to the basics."

In the past, children could leave school and get a job. This is no longer the case. Therefore, efforts to make school more appealing are appropriate. More definitions of school success are needed. The conventional practice of offering good grades only to high achievers is at odds with the societal emphasis on enabling everyone to have a favorable self-concept. The significance of self-concept has been misread by some teachers to mean that everyone deserves good marks. Consequently the inflation of school marks is rampant. At the same time schools are being criticized for low achievement, students are getting better report cards than any previous generation. In addition, it has become possible for almost anyone to attend college. In this kind of educational setting the identification of meaningful incentives for children can be a perplexing problem.

Because of our readiness to judge success on limited criteria, some authorities believe that grading should be eliminated, that school failure should be banned, or at least the criteria for success should be widened. Students also contend that it is a hateful thing to be forced to do something at which you fail over and over again. The recommended antidote is not eliminating grades or the occasional experience of failure. Rather, it is important to get down to the level of children's thoughts and find out what they do understand. This is done by asking questions and letting them talk. Based on the results, instruction can fit each individual's level of development.

Despite the years of discussion and experimentation on how to deal with differences in ability, no satisfying solution has been formulated. It seems that the major differences between pupils is ability or aptitude. These are particularly important because of their impact on motivation. Often those who begin a class at a competitively high level are motivated, while those who start at a low level are inhibited in participation. Two strategies are suggested: (1) make any rewards a group enterprise, and (2) let rewards be based on mastery rather than competition or time.

Peer Approval

Making rewards a group enterprise is a sound recommendation because in middle childhood peer approval can be a significant incentive. The motivation to belong and to be well thought of by peers becomes a powerful motivational factor at this stage. If teachers and parents can see this influence as being potentially useful, they can orient emerging groups to develop benign rather than destructive influences. This calls for observing groups in action and capitalizing on their interpersonal dynamics. Specifically, adults should seek action patterns that give those who are low academically a chance to exercise their strengths.

> Ted, a nine-year-old, was a poor reader and only average in arithmetic. He was the first one sought when games were to be played at recess. His athletic ability was truly amazing. A nine-year-old who could catch a frisbee between thumb and forefinger! A nine-year-old who could play baseball with the best of the junior high school boys! In grade school games he could hit a baseball out of the infield every time it was thrown close to home plate. When the ball was in the strike zone it was hit past the outfield. I saw him throw from third base to first and then back up the second baseman when the first baseman dropped the throw. He worked hard in class and spelling and arithmetic. He had the disturbing habit of discarding his waste paper in the waste basket with a jump shot over the shoulder. The playground supervisor was too busy to learn the Ted was a poor reader and too busy to tell the teacher what a prize athlete he was on the playground.

Parental Pressure

Because there are pressures from peers, community, and schools, parents need to search for ways to reduce the pressures generated at home. Part of this can be done on the basis of reflection. For example, most parents are proud of their children and brag about them with the child is not present. In the presence of the youngster, they are tempted to talk of potentials, making it seem that they are never quite satisfied. They keep prodding the children to be more skilled, get better grades, and to be socially superior to same age cousins and school mates. It seems that if those youngsters could just be two years older they would satisfy parental expectations and aspirations.

Part of the pressure reduction can be achieved by communication in which children express their feelings without condemnation or interruption.

> A working mother approached this by kissing and hugging her intermediate grade children. Then she would say something like this, "I know you have had your victories and troubles at school or with friends. But right now, let get my shoes off, get my feet up in the air, and have a Coke. Then at six o'clock I'll talk with each of you. When we get through we'll all pitch in and get supper." The children did just that. They quickly learned that Mom's first half hour at home was hers. The delay did not erase their eagerness to give their reports. At six o'clock, each in turn sat on the arm of her chair and talked of his or her day. Mother offered neither criticism of their behavior nor of the friends and teachers who might not have seemed all that blameless. She listened. On occasion, she asked for clarification.

Much can be accomplished also by discussion with other parents who are experiencing the same milieus and dilemmas; e.g., children under pressure from peers, parents, and school requirements.

LATER CHILDHOOD

Peers, Parents, and Planning

Peer approval tends to gain increasing significance during preadolescence. As boys and girls become stronger, wiser, and more autonomous, they must of necessity decrease the obvious parental power; yet they are not really sure they want to "go it alone." They turn to their peers for help and approval. However, it must be made clear that an increase in the need for peer approval does not decrease the power of parents. It is the obviousness that must diminish (Elkin & Handel, 1984).

Studies of the phenomenon of locus of control show that 10–12-year-olds continue to be influenced by how they have been , and are, treated by parents (Chandler, Wolf, Cook, & Dugovics, 1980). Locus of control refers to the degree to which persons feel responsible for what happens to them. Those with an internal locus of control (ILOC) believe that their skills, competence, intelligence, and power controls their future— their destiny. Those with an external locus of control

(ELOC) believe that luck, fate, chance, and other persons shape their present and future circumstances.

Parents of preadolescents who reveal an ILOC (assessed on the basis of observation and inventories) were helpful, encouraging, permissive, talked with their children, and intentionally encouraged independence. Parents of ELOC preadolescents tended to be authoritarian, exacting, to make decisions, and regulate their children's behavior. Once again, it becomes almost discouragingly apparent that growing children become a sad or glowing reflection of how they are dealt with by key persons (Eron & Peterson, 1982).

Guidance and Goal-setting as Motivators

Many educators believe that students can be better prepared for the world of the future than is the case at present. For example, teachers come to class armed with a set of objectives, usually expressed in terms of minimal competencies that everyone is expected to achieve. This is appropriate but in the quest to make schools accountable for basic instructional outcomes, the chance for students to become self-directed must not be eliminated. If they are to become accountable, autonomous, and responsible, the process of reasonable goal-setting must be given some exercise. They must be encouraged to make some of their own decisions. They must learn to select goals wisely, to modify their aspirations when necessary, and to free themselves from the unreasonable expectations of others. Teachers can facilitate goal-setting by encouraging students to try new activities without evaluative threat. Students must become acquainted with their limitations as well as their strengths. They need assistance in setting reasonable standards for performance.

Attempts to involve students in goal-setting have sporadically been made. Career education, initiated during the 1970s included student involvement in career exploration, participation in processes of decisionmaking, reflective activities over a period of time spanning several grades, and trying to develop awareness of the fact that careers will, through life, change and evolve. Here our concern is with the motivational power of preadolescents having a listened-to voice in setting personal goals.

It is recognized that without guidance, youngsters may make job choices which are not in their best interest—unrealistic in terms of the students' potentials and unrealistic in terms of job availability. For instance, in the city of Cincinnati, 1,658 girls and boys in 35 junior high schools were asked to state their occupational choices. Results of the inquiry were summarized by Super and Hall (1978) as follows:

> What would Cincinnati be like if these students became the sole inhabitants of the city in the jobs of their choice, ten years from now? . . . Health services would be very high, with every 18 people supporting one doctor . . . It may be, however, that they would all be needed in a city that has no garbage disposal workers, no laundry workers, no water supply personnel, since no one chooses to do that kind of work . . . The two bus drivers will find that their customers get tired of waiting and use the services of the 67 airline pilots. It may be difficult getting to Riverfront Stadium to see the 40 baseball players.

The emphasis should be on the process of selection, not on career choice. The role of guidance workers, teachers, and parents is to caution against the premature choice, choice by peer pressure, choice by parental pressure, and choice without thorough investigation. In addition, students should acquire knowledge about the necessity and adjustment of their having to make, during adult life, an average of five or six major job changes (National Center for Research in Vocational Education, 1986).

Adolescence

Boredom—Responsibility

One of the problems attributed to today's adolescents is boredom—the opposite of being motivated. Presumably because of the way our society is constituted and organized adolescents are, despite having rather mature bodies and minds, kept from performing adult roles. As a consequence, they suffer from the frustration of boredom, which in turn tends to trigger high rates of delinquency, pornography, drug abuse, theft, and sometimes violence. Government, business, families, and schools are seeking to counteract the boredom of adolescents. The search for ways to overcome boredom should continue.

During adolescence the motivational orientation changes, gradually and progressively, but certainly. Adolescents move toward the assumption of more responsibility. To be responsible means being accountable legally and morally for the discharge of a duty, trust or debt; capable of perceiving the distinctions of right and wrong; and able to meet obligations. History has demonstrated that the privilege of autonomy entails responsibility. This suggests that adolescents are answerable to home, school, and society at large. They must distinguish between right and wrong or society will make the distinction for them. With the aid of discussion and dialogue adolescents can progressively learn to meet more serious obligations.

Many parents consider the following responsibilities reasonable for teenage sons and daughters: appearing clean and on time for meals; caring for own clothes and possessions; doing schoolwork before pursuing leisure activities; doing homework and designated household chores regularly; and assuming some community role (such as visiting the elderly, helping in church activities, or tutoring young children).

This context for the development of motivation does not have to be forced on adolescents. With help, guidance, dialogue, discussion, and the opportunity for experience they will readily participate. Nora Richards, a Kansas City high school teacher, provided an elective class which focuses on service to others. Ninety percent of the seniors have chosen to work regularly as community volunteers in the past few years. Participants report ". . . new levels of awareness, a growing sense of themselves as important parts of the community, and a new understanding of individual needs and human dignity" (Shoup, 1980, p. 633).

The concept of responsibility can be applied to the handicap of boredom. The opposite of boredom is interest, stimulation, or involvement. The way to avoid boredom is to balance work and play—not be edict but by family discussion, vote or consensus, and practice. Sharing family chores might include shopping, planning and preparing meals, washing, mending, painting, house repairing, and servicing the car. Proper limits on televiewing are mutually decided and the decisionmaking often results in watching and discussing as a family. Keeping a diary of the good and beautiful things of life caused one girl to remark, "I don't see how anyone can find life dull."

Boredom occurs because we are in a monotonous environment or because we have acquired a frame of mind which leads us to conclude that a situation is boring. In the classroom educators can choose not to be responsible for the boredom of pupils. Instead they can decide to make pupils responsible by asking, "Why have you chosen to be bored? What other choices do you have? What are the consequences of the choice you made?"

Giftedness and Motivation

Being a gifted student can present difficulties in motivation. Perhaps this assertion seems strange. After all, isn't it true that teachers show favoritism toward students who have been identified as extraordinary? A less favorable impression of teachers comes from Mensa, the society of individuals who score in the top 2 percent of intelligence or achievement tests. Adult members of Mensa were surveyed about their elementary and high school experiences. Most of them marked "difficulty with teachers" as a frequent negative experience. It seems that many creative and talented students were scorned, ridiculed, and deprived of suitable challenge (Strom, 1983).

Still another perception is that teachers ignore the gifted and talented. Sometimes this occurs because of demands for accountability. In order to bring low achievers up to minimal competence, teachers feel obliged to spend much of the class time working with the low achievers. The tendency is to suppose that, so long as gifted children earn high grades and stay out of trouble, it is unnecessary to consider their dissatisfactions. By this reasoning we contribute to alienation. This would appear to be the case in Iowa where studies concluded that of all students with IQs over 130, 45 percent had grade averages lower than C. As was noted in chapter 3, the fact is that nearly 20 percent of all dropouts in the United States are gifted; this is a much larger proportion than their representation in the general population (Lyon, 1981).

Handicaps and Motivation

It is estimated that from 10 to 12 percent of the population have some type of handicap; this figure rises as screening becomes more pervasive and sophisticated. Appreciation of the overall potential of handicapped persons has been enhanced in recent years by greater public awareness regarding the performance possibilities of those who are blind. For many years blind persons were limited to craft occupations such as basket weaving and broommaking. Today, increased numbers of the blind are becoming business persons, professionals and high level technicians. As with other handicaps, segregation of the blind is being replaced by mainstreaming.

More than any other handicapped persons, the blind tend to lose interest in the environment. For normal persons, vision provides numerous stimuli; blind persons suffer a deprivation of stimulation that can lead to frustration. For this reason, blind children and adults may be more susceptible to frustration and hostility than other handicapped persons. However, as the range of their leisure activities expands the blind may experience a corresponding decrease in such negative feelings.

Recently, while millions of American fans were watching the Cardinals and Royals battle for the major league baseball championship, the authors attended a different World Series in Phoenix. There were uniformed players and supportive fans from Chicago, Houston, Minneapolis, El Paso, St. Paul, and Phoenix. All the teams had previously won local tournaments. They were playing for the championship of Beep Baseball, a fascinating game named for the sound emitted from the ball. What makes Beep Baseball unique is the close collaboration it requires of the players—most of whom are blind.

Each team has five blind fielders who are stationed throughout the regular baseball infield and outfield. These fielders are accompanied by two sighted observers called "spotters". The spotters' job is to position the fielders, warn them if a ball is hit directly at them, and retrieve balls that go beyond the ring of fielders. The batting team supplies its own sighted pitcher and catcher. Because they are teammates, the pitcher does not want the blind batter to strike out (five strikes are allowed). Rather, along with the catcher, the pitcher verbally encourages the batter to do well and tries to place the ball where it is most likely to be hit. Before each throw the pitcher is obliged to give two audible signals—"ready" and "pitch"—to signify that the ball is being released. Of course the batter can also hear the beep of the ball. Upon making a hit, the batter runs toward first or third base, where a battery-operated stand emits a loud buzzing sound. If the batter arrives at the base before someone on the opposing team finds and raises the beep ball in the air, a run is scored.

Because Beep Baseball players rely on hearing to locate the ball and to find the base, spectators must remain silent from the time the ball is pitched until the runner has been called safe or out by the umpire. Only then does the crowd cheer. Similarly, when planes fly over, the game is delayed until the noise subsides.

The motivation, skills and sportsmanship displayed by the Beep Baseball players are a delight to witness. Even fans who cannot see are enthused. As one blind spectator from St. Paul said: "Our team can do it, I'm sure. I came 1500 miles to see them win this game."

Motivation Based on Adolescent Needs

We can summarize the basic aspects of motivation during adolescence as follows:

- Adolescents have a need to be autonomous. This does not mean that they are free from social obligations as they choose curricula, careers, and companions. Their opinion deserves consideration.
- Adolescents in pursuit of autonomy need to have the support, sanction, and presence of peers. Adults can with considerable safety leave the choice of peers to adolescents by asking periodically, "What do you really want for yourself?"
- Motivation may be from external factors (praise, pay, privileges) or internal factors (healthy self-concept, long-term goals, search for esteem), but should move toward the latter type. The employer in adult life does not stand around looking for a chance to give praise.
- Motivation in adolescence must (because of the nature of social and psychological maturity) move toward concern for others; i.e., toward generativity.
- Motivation must derive in part from utilizing one's unique talents and from knowledge of progress in the development of those talents. Knowledge of progress will be self-evident in some instances (a gain in such skills as algebra, dancing, or job performance) and in others (social concern, communication skills) peers or adults must provide feedback.
- Motivation, at any age-stage, is enhanced by the process of involvement—having a stake in the goal, a feeling that one's presence and power makes a difference.
- By teaching, modeling, or pointing out evidence, the idea should be emphasized that adolescents can choose to be, or not to be, motivated in socially approved directions.

Independence, autonomy, and responsibility should be kept in mind as pervasive aspects of pupil motivation. Independence without responsibility is a delusion. If persons are to be independent, it is also necessary that they be accountable.

Early Adulthood

New Definitions of Success

Young adults are faced with a challenge of fusing their identity with that of others. If the fusion is successful, the person can live harmoniously with spouse, employer, fellow workers and neighbors. If the quest for intimacy fails the person is thrust toward isolation. Success on the job and in the home is threatened. Certainly we see a great deal of isolation in young families as they break apart in divorce, separation, or desertion.

Despite the prevalence of living together without marriage and the ostensible ease of separation, couples frequently experience a sense of failure when their relationship is terminated. Successful marriage continues to be a personal goal, a source of motivation and one aspect of resolving the developmental crisis of young adults searching for intimacy. In a nationwide life satisfaction poll, conducted by the Roper Survey Organization, 80 percent of the 2,000 participating couples identified marital happiness as one of their most important goals. Less than 55 percent felt their marriage was a happy one (Parnham, 1985).

Sexuality and Communication

Many people consider sexuality as a primary motivation. This is especially true after half a century of Freud and psychoanalytic psychology. Certainly at the peak years of sex prowess and sex interest, it is an important source of motivation. But with the emergence of a more liberal morality, the prospects for a healthy view of sexual relations appears to be diminishing. Teenagers, seemingly with the tacit approval of some parents, have lowered the age of sexual activity, increased its range, threatened its intimacy. In view of the high rates of unmarried pregnancies and the continued escalation of divorce and separation rates, it seems wise to associate sexuality with intimacy. This is what successfully married couples have done. Love and sexual intercourse are not synonymous—despite popular terminology. It is not claimed here that premarital sex and promiscuity make marital felicity impossible; but, promiscuity does make intimacy and success in marriage more difficult (Maddock, Neubeck, & Sussman, 1984).

Mature sexuality and successful marriage depend to a great extent on good communication. Glieberman (1981), a Chicago domestic relations lawyer, in answer to a question about causes of the nation's high divorce rate, said, "The No. 1 problem is the inability to talk honestly with each other." When courting, people put their best foot forward. They talk glibly about superficial things, but after marriage some are unable to talk. They cannot lay out a week's plan, discuss the budget, exchange thoughts about changing interests or emergent occupational plans. It is not economics, it is not sexual incompatibility, that cause marital failure. The most frequent cause of divorce is poor communication. This view seems to support Erikson's (1980) hypothesis that success for young adults resides in being mature enough to achieve intimacy.

Success, Intimacy, and a New Work Ethic

The definition of success in terms of sufficient maturity to achieve intimacy is apropos to work as well as to marriage. The usual criteria of business success are high salary, being the biggest producer of oil, autos, furniture, or cosmetics, high rate of dividends per dollar invested, and, sometimes, quality of product are signs of success. The United States was once the undisputed leader in this criteria but recent years have seen a decline as owners and managers take sides against workers in a perennial struggle to see which group has the greater greed. Workers seem to take too little interest in the quality of their product. Far from intimacy, the relationship between worker and owner is too often one of enmity and exploitation (Kilmann, 1985).

It can be different. Now and then an employer hires workers for the farm, factory, or store on what seems to be a family basis. Workers share the problems, share the work load, and share the profits. In one such case a box manufacturer's plant burned down. Insurance was insufficient to replace it. Workers shared the burden of replacement by going without wages. Competitors had enough respect for the box maker that they filled his orders during the rebuilding time. There are other similar reports of cooperative (as contrasted to competitive) factories, stores, and investment enterprises. Some workers are cooperative, empathic, and other-concerned. It seems, in short, that it is possible to achieve success on the basis of collaboration.

Japan has demonstrated that such collaboration can function on a nationwide basis. Ouchi (1981) calls the production plan which characterizes this cooperative relationship between owners and workers, "Theory Z." Quite simply, it postulates that involved workers are the key to increased and quality productivity. Many authorities are cited in Ouchi's book, but Erikson (1980) is not in the bibliography or index. Hence, some quotations are of interest: "The common thread in Japanese life is intimacy. The caring, the support, and the disciplined unselfishness which make life possible come through close social relations . . . In contemporary America there is apparently the idea that intimacy should only be supplied from certain sources" (Ouchi, 1981, p. 8). He mentions the family and the church as the traditional sources of intimacy in the United States. Business and industry are not cited. As noted, intimacy can be difficult to achieve in the family; but it is not an impossibility even in work.

With this kind of interpersonal orientation the Japanese have been able to maintain a viable work ethic. Instead of "Thank God it's Friday" workers ascribe their feelings as being a member of the employer's family. A genuine source of shame would come from being disloyal to the employer. In contrast, Americans are more inclined to feel loyalty only to themselves. They have ". . . become soft, lazy, and feel entitled to the good life without earning it" (Ouchi, 1981, p. 11).

In this section we have slanted the discussion of motivation to what might be more than to what now exists. Because early adulthood is a time of strength and decision making, we recommend adopting a proactive orientation. Three developmental areas were used to illustrate the potential of planned perspective. It is admitted that the pleasures of sex are intense; but they are transitory. It is admitted that intimacy is difficult to achieve; but the sadness of isolation is real. The achievement of high salary and prestige accords with current mores; but the dividends from cooperative endeavor is more in accord with the basic meaning of being truly human. It seems early adulthood is an appropriate time to reflect on the motivational power of intimacy in work, marriage, and lifestyle.

Avoiding Narcissism

Narcissism, a word derived from Greek mythology, means excessive self-admiration or self-love. This love of self is so great that it tends to exclude responsibility and concern for others. It is a defeat of intimacy. The handicap of narcissism is not limited to young adults, Hollywood beauties, or outstanding athletes; but it most frequently occurs in beautiful and especially talented persons. Psychotherapists find narcissism to be a highly significant diagnostic category of psychological and social immaturity.

Adults can choose to abandon the lifestyle of self-love, self-absorption, excessive self-concern, and self-indulgence that would generate or perpetuate narcissistic tendencies. They can choose to pursue a course of concern for others. This is the path that leads toward generativity. Making the decision might involve such dilemmas as the following:

> How can I find the right person to marry? *versus* How can I become the right mate?
>
> How can I find the job that fits my interests? *versus* How can I develop interest in the job available where I am?
>
> What can I do to enjoy my leisure time? *versus* What social role will allow me to make a contribution to the community?

The study of narcissism, for the most part, focuses on individual behavior. Wallach & Wallach (1985) say that we should be concerned about what social practices and institutions produce or counteract it. This is precisely what motivational maturity really means—it involves moving into the community, schools, churches, and government to make a constructive influence.

MIDDLE ADULTHOOD

The Empty-Nest Syndrome

The empty-nest stage refers mainly to women in midlife who suddenly waken to the fact that their mothering and guidance of children is no longer needed or desired. Certainly not all, or even a majority, of women suffer this syndrome; and a few men lose their purpose when the responsibility for childrearing is over. Adolescents have "flown the coop" and with varying degrees of emphasis make it known that they are now free to do their own thing. (Incidentally parents should be proud of having done their job when this independence is asserted.) The mother broods on the fact that her purpose in life have been fulfilled and that she will just be "marking time" for the rest of her days.

Such feelings may have had a better reason for being at an earlier time in history. Few women then lived much beyond childrearing days. Now many more work outside the home than formerly. They can go right on with their work. Others can and do find acceptable substitutes for their earlier emphasis on mothering. In recent years the avenues of expressive lifestyles for women have proliferated. Moreover, because husbands also may be undergoing a midlife crisis, wives have the challenge to being other-oriented in their efforts to improve the marriage relationship despite its changed function and values (Eisenberg & Eisenberg, 1985).

Because there are so many ways to avoid the psychological hazards associated with the empty nest, we propose no specific alternatives. Books, magazines, and talk shows provide plenty of good advice. We do suggest that women join discussion groups that deal with personal problems. In this way the particular variety of distress that middle-aged women experience as motivational problems may be dealt with diagnostically.

And the Not-So-Empty Nest

The influence of cultural situations on human development is illustrated in the current reversal of the empty-nest syndrome. Young adults in the depression years of the 1930s left home to avoid being burdens on their unemployed parents. They had been brought up with an emphasis on personal discipline and responsibility. More recent generations growing up in an era of children's rights now, as adults, return to their parents and reoccupy the empty nest.

Silden (1982) reports that the economic squeeze, rising costs of housing and education, and instability of marriage are causing many young adults to return to their childhood homes for protection. Prospective doctors, lawyers, and business persons reside at home to cut costs while they continue their education. Young women with rising frequency return home with a baby, and sometimes a husband, to thrust middle-agers again into parenting young children. However much mothers have regretted their children growing up, most are not prepared to live in one household with adults who are now wage earners and/or parents. Some authors contend that today a midlife crisis is almost inevitable; i.e., the need for making a major change in lifestyle. Those who accept the fact of changing life patterns at each stage of life have the best chance of success and satisfaction (Nichols, 1986).

Menopause and Climacteric

Menopause is related to motivation. An estimated 15 percent of women experience enough physical discomfort to require medical attention. Some women, being forwarned, take the menopause in stride and report minimal disturbance. Others experience a level of emotional distress which exceeds the basic physical discomfort. Representative of the turmoil and anxiety are such worries as these:

> Children moving from home and leaving the mother without purpose.
> Elderly parents becoming more dependent and creating financial pressure or crowded living.
> Fear that the spouse no longer is sexually attractive and one is becoming less interesting.
> Fear that one's social value declines with wrinkles, sagging facial muscles, and gray hair.
> Mood swings cause anxiety about one's emotional stability—"I must be crazy to worry about these things."
> Fear that the boring daily routine will continue until death.

Many persons in midlife are not greatly bothered by these issues. For those who are concerned, it is important to know that they are not inherent in the menopause. Nicholson (1980) on the basis of his Colchester (England) Study of Aging, says that the concept of age-related crises is not based on fact. A more positive approach is to examine one's lifestyle. Positive steps in motivation can be encouraged and strengthened by discussion with others who are going, or have gone, through this natural passage from one era to another. The "support-group approach" helps women understand the normality of the whole process. It enables them to see that they are not so different from other women as they thought; they can develop new goals and activities; and it is appropriate to reevaluate their relations with their husband. Here are some discoveries reported by participants in support groups: I learned that state of mind has much to do with well-being...I decided to talk with my husband about my feelings...I started swimming classes and walking tours...I took an aerobics class...I learned that many people are living longer and facing the processes of aging...I discussed my mood swings (from depression to enthusiasm) with my husband...I abandoned my earlier thought about menopause being an end and began thinking of it as a new challenge.

Men also experience physiological changes that affect sex life, goal changes, and motivation. Some of these are due to psychological distortions. This change, as discussed in Chapter 2, is called the climacteric; and sometimes referred to as "male menopause." Men also can feel that their usefulness as a breadwinner, adviser, and confidant of those who are younger has declined because children have assumed adult roles. American society places great emphasis on youth, vigor, beauty, and sexual attractiveness. Men begin to notice that their hair is thinning or getting gray. They are putting on weight, especially around the waist. Worst of all, they notice a decline in sexual virility and interest. This decline, including a diminished sperm count, is a normal accompaniment of the climateric. In time, it is evident that some men are unable to have an erection or always maintain it during intercourse.

The climacteric occurs somewhat later in men than the menopause does in women. For some it may begin in the late forties, with a small percentage not reaching it until about the age of seventy. This decline has actually been going on since about age 20–24 but the change was so slow as to go unnoticed. The children leaving, the realization that his wife is not so glamorous as twenty years ago, the fact that his climb up the occupational ladder is not so rapid (or may even have ceased) all combine to create, in some men, a mid-life crisis. But the crisis consists largely in outlook rather than being an essential part of organic change (Maddock, Neubeck, & Sussman, 1984).

Psychological studies which have been made on the middle years lead to the conclusion that it can be a time of changed motives and solid satisfactions. Neugarten (1980) refers to middle age as the new prime time of life. It is realistic to acknowledge that many middle-agers move into a new, productive, and gratifying life stage. Because of improved health and prolonged vigor as compared with the beginning of the century, "...An enormous number of people are starting new families, new jobs, and new avocations when they are 40, 50, or 60 years old" (Neugarten, 1980, p. 3).

LATER ADULTHOOD

Until recently the common conception of old age was rather bleak. The elderly were pictured as poverty-stricken, encumbered with health problems, and as experiencing marked cognitive decline. This gloomy and erroneous picture was probably due in considerable degree to the fact that pioneer studies of the aged were made on institutionalized populations—captive audiences. Hence, those persons who were in homes for the aged, rest homes, and mental institutions were the unfortunates who provided the data for this one-sided view. While geriatrics (medical aspects of aging) and gerontology (the psychology and sociology of aging) have made great forward strides in the past two decades, the complete picture includes the fact that some aspects of aging are unpleasant and discouraging. Many people experience decline in hearing, strength, vision, agility, and digestive and circulatory efficiency. Discouraging factors include weakened family ties, loss of friends, and moving away from the long-familiar neighborhood. Because we are all part of the human family, positive motivation needs to be sought by the younger generations, and the healthy elders to develop compassion for the less fortunate older persons.

The lonely and less healthy old person is not the total picture either. Many in retirement and well past the age of retirement are healthy, vigorous, and love to travel, avidly pursue their avocations, enjoy their sexuality, responsibly engage in social service and otherwise are positively motivated. "We are maturing as a nation as we slowly overcome the anxiety about the ongoing process of growing older and begin to see later life as a viable, vital stage with exciting possibilities of its own" (Lewis & Butler, 1985, p. 60).

Imaging and Self-Assessment

Imaging is one productive approach to developing age-appropriate motivation in later adulthood. In imaging one is involved in stretching the imagination, thinking of new ways to use familiar tools, and to verbalize imaginary ventures and travels. In recent years imaging has expanded beyond art and creativity to physical and psychological therapy, personality change, marital serenity, occupational aspiration, and social compatibility. Some medical doctors, having observed that certain patients die because of seemingly minor illnesses, have begun actively to search for better understanding of the mind-body relationships. One experiment involved imaging forces (e.g., knights on white horses slaying enemy cancer cells) that counteract illness (Scarf, 1980).

The most noted publicist of imaging is a clergyman, Norman Vincent Peale (1982). He recalls that as a young minister he became discouraged about the difficulty of filling the pews of his Fifth Avenue Church in New York City. Later he had a hand-to-mouth existence when he started publication of the periodical, *Guideposts*. A woman on the staff told him he was thinking 'deficit.' "You must start imaging prosperity instead." When the thoughts (the imaging) become positive his actions became more confident and the confident air influenced his associates. His positive thinking approach has since been adopted by millions of people. Older adults need periodically to pause and examine their sense of possibilities.

Isolation, Communication, and Motivation

Psychologically the great hazard of old age is isolation. This same condition, isolation, in childhood is regarded as a danger because it has a negative effect on cognitive, social, and emotional development. Isolation is, in part, a function of society; it seems easier to put old persons aside (in a room, community, or institution) than to assist them in their efforts to make meaningful cross-age contacts. In part, isolation occurs because older persons expect others to initiate contacts. These older persons seem to have accepted certain misconceptions about the accompaniments of aging which causes them to doubt themselves. Some of these misconceptions, as enumerated by Pearson and Shaw (1982) are:

Life extension would cause overpopulation.

Life extension is selfishly motivated.

Old persons cause social stagnation.

Family structure is threatened by aged persons.

Old persons and younger persons have intrinsically based contrasting life purposes.

Extended life is boring.

Extended life uses up the capacity of the brain to store new experiences.

To the degree that any of these notions contain an element of truth it is because of the attitudes that are taken toward them by older persons.

The opposite condition of isolation is involvement. When reference was made to isolation it was stated to be a hazard. This does not mean that isolation is a condition of the majority. No studies that we are aware of reveal whether isolation or involvement is more prevalent. Some persons (by inclination, lifelong habit, deliberate choice, or by peer influence) remain involved. Others, by choice or concentration on personal interests live in robust mental health with what appears to be isolation. In fact, Maslow (1970) suggests that many self-actualizers have especially deep ties with "rather few individuals" (p. 166).

Each individual should seek whatever means of involvement brings him/her the most personal satisfaction. One way to avoid some of the problems of old age is to study them in advance of their occurrence. The search for positive motivation may be initiated in discussion groups devoted to the problems of elderly persons. Such discussion groups are provided in many churches, community colleges, centers for senior adults, and where not already in operation they can be readily formed because there is a substantial number of people in the top most age-stage of development. Programs that include intergenerational communication are even more beneficial.

Grandparenting

The conventional and obvious motivation for the elderly is being an effective grandparent. In the past decade this role has come to be regarded as a motivational factor in both the development of the young and in the optimum adjustment of the elderly. Tice (1985) has described programs which involve older people as helpers in nurseries and preschools and adolescents as companions to the elderly handicapped and lonely. Whitley and Duncan (1985) report that after adopted grandparents visit children a few times, social motives develop which are closer than kinship ties. In grandperson programs where children from daycare centers visit the elderly (both sick and well) there is benefit for both age groups. The young learn about the reality of aging and the elderly experience a ray of sunshine.

The problems of grandparenting are often confusing and sometimes difficult (Bengston & Robertson, 1985; Kornhaber, 1986). There are few precedents because of such variable factors as culture, distance, and time. Culture: Countries and families differ in the incidence of nuclear and extended families. Distance: Families separated by distance have different problems of communication than those living within walking or short-drive range. Time: Families which have kept in touch have different relationships than those who visit only once a year (or five years). But where grandparenting is active, salutary motives are generated for children, parents, youth, middle-agers, and the elderly.

Chapter 7

COPING WITH STRESS

There is good news and bad news regarding the nature of stress and its management. The bad news is that the more studies we have about stress the more complex the phenomenon becomes. In the past stress was less frequently mentioned and could be reduced, allegedly, by weighing the situation, avoiding tension when possible, and then just taking it easy. Persons were advised to stop trying to accomplish difficult goals overnight, take care of physical health, and be congenial at home, school, and work. One would then live longer, be more productive, and experience greater satisfaction. But stress is just not that simple. Stress may occur before one learns to think logically and it combines a wide variety—a life full—of events and people.

Stress means different things to different individuals. Physical differences between people begin to express themselves even before birth and so does stress. Some babies are victims of fetal stress and trauma during birth. One infant may be born serene and happy, with enthusiasm and vigor. Another baby born into the same family (no matter whether earlier or later than the happy one) may be apathetic, and have poor appetite, digestion, and elimination. In short, differences in prenatal conditions, hereditary potential, birth experiences, and parental care all have varying effects on stress. Subsequently, school teachers, work conditions, supervisors, occupational influences, workmates, spouses, parenting demands, in-laws, sickness, accidents, and world conditions paint different strokes on the picture of stress.

The good news is that, along with the growing recognition of the complexity of stress, research is providing more and more clues regarding how inevitable stress can be more competently handled. And because of these helpful clues, practical programs of prevention and tolerance can be better adapted to individual requirements. Instead of the broad generalization "take it easy," persons at most ages may be taught specifics on how to deal with stress. Our focus in this chapter is to (1) explore the meanings of stress; (2) identify some of its causes and signs at various ages; (3) consider ways to reduce pressure on others; and (4) learn methods of managing personal stress.

THE CONCEPT OF STRESS

Meanings of Stress

Stress refers to a condition of strain, pressure, or urgent adversity and may be manifested either physically or psychologically. It causes the heart to beat more quickly, breathing becomes more rapid, digestion slows, perspiration increases, muscles tense, pupils dilate, and stored fuels enter the blood. One gets ready for fight or flight.

There has been a growing interest in stress since the term was first introduced by Selye (1956). Stress is now being perceived in terms of unique individuals as they interact with life events and people. The widespread concern about coping with pressure was reflected by a study in which 82 percent of the participating adolescents and adults reported feeling a need to reduce stress in their daily lives (Yankelovich, 1979). The American Academy of Family Physicians reported that two-thirds of office visits to doctors are prompted by stress-related symptoms (Wallis, 1983).

Some people seem to equate stress with tension and motivation. It is regarded as the universal urge required to deal with the problems of staying alive, getting enough to eat, and dealing with the unsolvable problem of death. Another view defines stress as a ravaging force that incapacitates the mind and body. This impression must be avoided whenever possible because it may keep one from functioning effectively in either normal day-to-day functions or in crisis situations. Some stress is coexistent with life itself.

There is little advantage in equating stress with either motivation or incapacitating pressure. Instead we prefer to regard it as strain that disturbs optimum functioning of the organism. Stress can include cold, heat, noise, pollution, food or water deprivation, and such psychological things as frustration, conflict, and discomfort which activate defense mechanisms or coping strategies. This is the predominant concept of stress. It is a vague definition, and must remain so, because for different people the stress point widely varies. A high school basketball player—or a test taker in a trigonometry class—must be "up" for the event. However, if the felt burden is too great, performance will be grossly impaired. Much will depend on the coping skills. It is not just the weight of external stress but *how it is viewed* that is the crucial consideration (Goleman, 1986).

Varied Vulnerability

Research shows that personality differences are reliable clues to performance under stress. Kobasa (1984) uses the concept of personality types to compare stress vulnerability. At one extreme are those who are stress resistant; they can carry a heavy load of stress and still function effectively. At the other extreme is the personality type which begins to fold under slight pressure. Stress resistant persons are those who reveal distinct and often measurable personality characteristics. They are objective and can distinguish themselves from external milieu. They are open to change and may in fact welcome change as a challenge to growth. They have a feeling of involvement; they feel that what they do and say makes a difference. They have a sense of control over their lives. They are inner-directed—they have an internal locus of control. The contrasting personality type is just the opposite. They have what is called an external locus of control—a belief that luck, fate, and other people control them.

What this adds up to is that how a persons looks at things is even more important than what is there. A loud-speaking supervisor may for instance produce all the physical disturbances of anxiety in one person while another says, "Both the supervisor and I have the same goal—improving effectiveness." Those who are open to change, feel

Physiology and Drugs

In addition to presence of stress factors and personal perspective, physiology and drugs also play a part in stress. Years ago it was discovered that a substance called adrenalin could produce an apparent state of stress in animals. A cat injected with the hormone adrenalin could be made to show fear or rage even in the absence of external stimuli (e.g., a snarling, teeth-showing dog). Similarly, humans injected with adrenalin can be made to feel readiness for anger, fear, hate. Small provocation may precipitate the emotion. Drugs can also be used to calm persons. In our anxiety-ridden world physicians often give prescriptions which have the effect of making persons more placid. Periodically we hear about entertainers and professional athletes who take drugs in order to feel high. After the performance they have to take sedatives in order to get some sleep and rest before the next session.

Blood lactate (a normal by-product of vigorous exercise) levels are a concomitant of anxiety. Pitts (1969) found that neurotic, guilt-ridden, anxious people had high levels of lactic acid. Lactate is a product of cell metabolism and rises under physical exercise or stress. Feelings of anxiety can be heightened by exercise or be induced by injections of lactate. The cause and effects are not clear. It seems that chronically anxious people produce more lactate than stable persons when doing the same amount of exercise. What is clear, from such studies, is that there is a relationship between body chemistry and psychological functioning. However, drugs, chemistry, and exercise alone will not do the job of maintaining balance, motivation, or perspective.

We come again to individual differences and what is known as the paradoxical effects of drugs, i.e., the same drug that stirs some people out of apathy often has a calming effect on others. The same alcohol may cause one man to fall asleep at a party, another to be happily congenial, and still another to be disgustingly belligerent. Each human body has a delicate chemical balance. It is dangerous for persons to experiment with drugs because of the paradoxical effects and the unique chemical balance, sensitivities or allergies. Yet young people as well as older ones who are misled by advertising resort frequently to self-prescribed drugs (Bezold, 1985).

Coping and Stress Reduction

Stress has a cumulative effect. The existence of stress may be tolerable for a time but finally wears one down. Children too may suffer cumulative stress from persistent and unrealistic parental expectations. Later, in middle age, those same persons may experience the cumulative stress that comes from taking care of aging parents. In a similar manner, some of the techniques for coping, or stress reduction, are applicable, in some form or another at several age-stages. For example, group discussion, meditation, relaxation, and temporary withdrawal are discussed in the following sections in regard to particular ages but are pertinent to other ages as well. Minor modifications of techniques may be necessary.

INFANCY

The Needs of Infants

The most basic needs of infants are physiological: food, water, warmth (appropriate bedding and clothing), exercise, and rest. While these needs are rather well met in technological societies, they are unmet in many of the developing countries. One estimate is that half the children in the world go to bed hungry. Their pain of hunger is an effective producer of stress.

After physiological needs are met, or at least on the way to being met, infants have safety needs. This means that they require protection from disease and injury; and they require the psychological comfort of having dependable, readily available caretakers. Babies whose mothers have to be absent for several days or weeks show their stress by becoming listless, unresponsive, and by refusing to eat. Stated briefly, a safety need of infants and babies is continuity of parenting. According to Maslow (1970, p. 40) the infant "...seems to want a predictable, lawful, orderly world. For instance, injustice, unfairness, or inconsistency in the parents seems to make the child feel anxious and unstable."

> Sara came to the psychiatric clinic because she had voluntarily given up her infant son when she had, in rage, beaten him "two or three times." The boy was illegitimate but despite steady haranguing by Sara's mother, Sara had tried to care for the infant

herself. The burden of child care plus work was too much and she had lost control when the baby cried. Now, Sara said, she wanted to "get my head on straight so I can get my boy back and be a good mother."

Maternal stress does not have to reach the stage of battering to do harm to the infant. Consistent rejection is enough to create accummulative overload of stress. Fortunately, there are a growing number of centers where parental victims of stress can get help and advice. For example, one innovative plan takes places at the extended family care center, originally developed in San Francisco. At this "home away from home," isolated parents are offered the resources of an extended family. The service that abusive parents need the most but receive the least is periodic relief from child care. To meet this need, the center offers free day care.

The use of day care for abused children is a new and promising alternative. From the parents' standpoint, the chance to do something without the children and yet be sure they are being looked after is most gratifying. For the professional staff, day care is seen as a protective safeguard as well as an opportunity to teach children about successful relationships. At first the children typically distrust the teachers, abuse younger peers, and demonstrate that they don't know how to set limits on their behavior. Limit setting is a crucial but difficult objective for abused children to achieve, because they come from families that demand unquestioning obedience in a manner that prevents children from learning to control themselves. In time most of these boys and girls develop a confidence in their teachers, become more able to see worthwhileness in other people, and begin to develop the sense of trust they will need to establish intimate relationships.

A second objective of the center is to develop the parents' reliance on one another. To appreciate the benefits of this extended family, we must bear in mind that abuse is episodic and usually occurs during times of family stress. Money problems, work dissatisfaction, even something as unrelated to the child as the failure of a kitchen appliance can precipitate abuse. As a rule children become the target for violence because parents do not perceive any other outlet. However, if parents can call each other, talk when they are angry, and feel comfortable about disclosing negative emotions, their aggression is redirected and loses is destructiveness. In order to help parents become more open about problems and better able to share feelings with other members of the extended family, two special consultants are on the staff. What makes this pair special is that they were once abusive parents themselves. These consultants fulfill a vital function in developing trust and communication between the parents and staff (Gelles & Cornell, 1985).

Symptoms of Stress

Loss of appetite, uneasy sleeping patterns, frequent illness, and temper trantrums are the obvious and common symptoms of infant stress. These irritating symptoms tend to make it more difficult for caretakers to be predisposed to making efforts to supply the deficiency. The most serious symptoms are withdrawal, apparent resignation, depression, and silence. The infant it seems has given up fighting and because he cannot physically retreat, he does so psychologically.

In contrast, infants who experience warm, consistent, and loving care soon develop a sense of trust. They readily permit their caretakers to get out of sight because they know they will return. They respond to others in their household who are not major caretakers. They are demonstrating that they have taken the first step in the continuing process of socialization. It should be noted that caretakers need not be natural parents. Foster parents or substitute parents can provide the safety need—as long as they are consistently reliable—and thus reduce stress (White, 1985).

Threat to Attachment

Bowlby (1980) summarized studies on attachment related to infant stress as follows:

> Prolonged separation from the mother constitutes stress for the infant once the attachment has begun.
>
> After separation attachment manifestations tend to be intensified
>
> Minor short separations may in some infants create stress.
>
> Maternal fatigue, depression, or preoccupation with other concerns may contribute to the insensitivity which leads to infant stress.

Earlier investigations of nonattachment were referred to as maternal deprivation and failure to thrive. Studies of orphans, abandoned babies, hospitalized infants, and other mother-absent infants showed common symptoms of stress. Institutionalized babies, despite good nutrition, medical care, hygienic conditions, and prescribed schedules but who lacked adequate human contact showed frequent instances of failure to thrive. By one year of age these deprived infants showed markedly low mental development. Their death rate in a measles epidemic was one in four compared to one in 200 with home-reared infants. It is believed that disrupted mothering causes many psychological and social problems in western culture. Disrupted mothering includes mothers who are physically there but psychologically absent, a succession of mother substitutes, and in sharp contrast, "smothering"—not allowing the infant any time by itself to discover its own powers (Klaus & Kennell, 1983).

Maternal deprivation, failure to thrive, child abuse, and child neglect have consequences that reach into adolescence and adulthood. The infant victims are at high risk for becoming a-human, at best. At worst, there is high risk for their becoming inhuman psychopaths and sociopaths. All of us have read reports indicating that someone has committed an unbelievable crime against children, adolescents, or adults, and when caught show no remorse, grief, or guilt. These persons have certain characteristics in common. They showed no emotion. They seemed to lack feelings and conscience. Two decades ago, the initial reporter of this phenomenon warned there is no simple solution, "But to a very large extent, the disease of nonattachment can be eradicated at the very source, by ensuring stable human partnerships for every baby." (Fraiberg, 1967, p. 57).

EARLY CHILDHOOD

Aggression and Withdrawal

The symptoms of stress in childhood that have over the decades claimed major attention are overt acts such as stealing, aggressiveness, destruction of property, and other antisocial behavior. Mental health professionals have been telling us for the past three or four decades that these are not the most important symptoms of stress.

Such overt behaviors, irksome as they are, inform us that the child perpetrators are under stress—they need help. They are threatened, overburdened, and underpowered but they are still fighting, struggling, and trying. With understanding help they may overcome the source of stress.

The more significant symptoms of stress are those of retreat, withdrawal, depression, and resignation. The individual showing these behaviors has quit. Some authorities believe that depression is the most serious malady of our time—we have moved from an age of anxiety to an era of depression. It is not just a matter of persons living in difficult situations but that persons have given up hope. Depression is a disruption of attachment bonds. And such bonds are essential to the survival and development of our species.

Most persons have read about, but may not have made serious note of it, that withdrawal and aloneness are frequently cited in spectacular news. To illustrate, Lee Harvey Oswald, John Hinckley and Mark Chapman were each known as loners since childhood (McMillan, 1982). Their desperate grasp for attachment, attention, and recognition contributed to the assassination attempts involving John F. Kennedy, Ronald Reagan, and John Lennon. Some of the most bizarre murders have been committed by loners. It is because of the possibility of loners becoming psychopaths that psychologists have recommended clinical attention to the shy, quiet, withdrawn children who are in no social circle. The aggressive, acting-out children will call attention to themselves. The recluse, the nonattached child might easily avoid attention at the very time that help might be of maximum benefit.

Ethnicity and Stress

Growing up is generally more stressful for black children than for white children. Even after two decades of vigorous civil rights and economic legislation, the environmental conditions remain inequitable. Stated in terms of risk, the Children's Defense Fund (1985) indicates that compared to white children, black children are two times as likely to die in the first year of life, be born prematurely, suffer low birth weight, have mothers who received late or no prenatal care, be born to a teenage or single-parent family, and live in substandard housing. They are three times as likely to be poor, have their mothers die in childbirth, stay with a parent who has separated, live in a female-headed family, be placed in an educable mentally retarded class, or die of known child abuse. Black children are four times as likely to live with neither parent, be supervised by a child welfare agency or be murdered before 1 year of age. They are five times as likely to be dependent on welfare, become pregnant as teenagers and twelve times as likely to live with a parent who never married.

Intimidation by Peers

Our society is very concerned about the mistreatment of some children by their parents. At the same time as we try to eliminate violence within families, perhaps we can become responsive to the more common experience of child abuse by peers. As an example, consider the situation at school bus stops. Urban living and school integration have combined to require an increase in busing. At suburban and inner-city bus stops, children of varying ages—often bored and wanting excitement—daily congregate without adults present. This unsupervised situation promotes bully behavior.

Most parents know the uncertainty that arises when children allege harm at the hands of peers. Depending on the number of physical size of the child's opponents, we are ambivalent about what to do. We are torn between the desire to protect our child and the recognition that we cannot always be bodyguards. Some parents decide that the only alternative is to train the young to protect themselves. Usually this means exposure to some form of judo, karate, or father-taught boxing lessons. An opposite response is to encourage the child to never fight back. Neither of these strategies seems to produce the desired effect. The young boxer often finds that his opponent has older friends who are prepared to avenge the defeat. Meanwhile, the unresponsive, nonviolent child learns that refusing to fight back leads to being called a sissy, which in turn invites even more people to bully him.

There must be better ways to eliminate the mistreatment of children by peers. To begin with, we ought not be deflected by the argument about permitting kids to abuse each other because "That's the way the world is"—unless in fact we feel the world should remain that way. It seems ironic that we even urge the young to fight their own battles when grownups no longer accept such advice for themselves. Many of the people who live in apartments and retirement communities have adopted the security guard as a necessity. An increasing number of adults are willing to pay for protection so that they can feel secure, so that they will not be bullied. No one should be expected to suffer intimidation, least of all the most defenseless—the very young. If protection is needed, the rotation of parents at the bus stop is one possible solution to the problem.

Unless society also expresses concern for the rehabilitation of young bullies, their future is in jeopardy. Researchers at the University of Chicago followed the careers of 875 third graders for more than 20 years. The results show that the patterns for aggressive behavior are set before age eight. One in four of the physically aggressive boys, those who pushed, shoved and took belongings from their classmates, had a criminal record by age 30. The comparable risk factor of other children is 1 in 20. As adults the bully boys tended to abuse their wives and act like demons on the highway. The women who bullied peers during childhood were inclined to severly punish their own children (Huessman, Eron, Lefkowitz, & Waldor, 1984).

Preschool and elementary teachers should acknowledge their responsibility to help develop personal relations skills and in turn take the bully's problems just as seriously as if the child had some other disability that adults feel more comfortable in treating. In the final analysis, when parents, teachers and caretakers of young children do not reject violence and teach boys and girls to do the same, they inadvertently support its continued expression.

MIDDLE CHILDHOOD

Divorce and Separation

A prime producer of stress in middle childhood is divorce or separation of parents. While broken homes do produce stress in early childhood, young children are less cognitively aware of divorce as the source of stress than are older children. Older children miss the parent who is absent, perceive the anguish of adults, and all too often suppose that they themselves caused the break. Fortunately, with help they can learn to adjust. That more of them will have to do so is quite clear. The proportion of single parent families (26 percent) has more than doubled in the past two decades. The causal factors have changed as well. Whereas widowhood used to be the major factor, divorce and separation presently accounts for two-thirds of all single parent arrangements. If the current trend continues, it is estimated that half of all the children now being born will live in a single parent home. Some of those children will enter a second family when their parent remarries. Unfortunately, the odds are even that this relationship will also end in divorce (Bumpass, 1984).

Parent Cooperation

In order to adjust, children of divorce need help from adults, especially their parents. It is important that boys and girls can continue to have a good relationship with both mother and father. This means they should be able to count on regular phone calls and visits with the noncustodial loved one. Neither parent should make unkind remarks about the other to the child or require a declaration of loyalty. It is also essential that children in single parent homes experience the sense of security that goes with adequate income. In 90 percent of divorces mothers are assigned custody. But many of them have not received the court ordered child support payments from fathers. For this reason inadequate income is regarded as the major problem of single parents. Lacking sufficient finances, such families often have to change their residence and many must apply for welfare assistance. In 1983 the rising cost of child support default to society prompted the government to intervene. Instead of relying on reciprocal state agreements to garnish the salary of noncustodial parents, federal authorities are now empowered to do so. Enforcement is making poverty less common among single parent families (Feldman & Feldman, 1985).

School Support

At school the child of divorce may or may not be a behavior problem, lack academic motivation or be withdrawn. But teachers often wonder what they can do for students who are undergoing domestic shifts, boys and girls who miss an absent parent and feel helpless because they can do nothing about it. The temporary emotional disturbance of such children is predictable, and teachers can help most by modifying their expectations. This change may take the form of shorter or less frequent homework assignments or longer time limits. What matters most is that no child involved in a troubled home situation be expected to perform as though conditions were normal.

When children exhibit signs of worry, anger, resignation, or the inability to get along with peers, teachers should meet with the school counselor and custodial parent to discuss the worthwhileness of a support group. Four goals are usually pursued in support groups. One is to let the child know he is not alone. Loneliness is one of the most difficult problems of divorce. According to Wallerstein and Kelly (1980, p. 70) who studied 60 families following divorce, "We were struck by the high incidence of intense loneliness that we observed in 27 percent of the children." They did not like coming from school to an empty house or having no family activity on weekends.

Because the child often feels miserable and deprived, the provision of pleasure is a second goal. Snacks and a friendly environment can make the group experience enjoyable for those who otherwise feel trapped by depression. Bear in mind that divorce is the single greatest cause of childhood depression. Wallerstein (1984) estimates that perhaps 70 percent of the clients seen by child guidance clinics are from nonintact homes. A review was made by the family backgrounds of several hundred children who had been treated consecutively for evaluation by the Department of Psychiatry at the University of Michigan. It reveals that children of divorce showed up at the clinic at twice the rate of their occurrence in the general population (Packard, 1983).

A third goal of support groups is to be with an empathic adult who uses active listening to let the children know he cares. The final goal is to center small group discussion on day-to-day issues which children must face. For example, what can I do when I'm angry or worried that will make me feel better? How can visitations be better? What things can I influence and which ones are beyond my control? Overall, children in support groups feel better when they know what is going on, when they feel understood and when life's difficulties are interrupted by a periodic relief and pleasure.

The Personnel and Guidance Association of North Carolina decided to assess the effects of counselor-led groups on children from divorced families. The participating third through sixth graders throughout the state were assigned randomly to an experimental or control group. Intervention consisted of eight sessions with a focus on practicing self-disclosure, discussing feelings, brainstorming, describing difficulties and/or advantages of divorce, and listening to compliments of others in the group. The results showed that the experimental group, but not the control group, made statistically significant changes in improved attitudes toward divorce and improved class conduct grades. It appears that elementary school counselors can have a beneficial effect on children's adjustment to family change (Anderson, Kinney, & Gerler, 1984).

LATER CHILDHOOD

Changing Nature of Stress

One advantage of dealing with all age-stages in each chapter of this book is to emphasize the changing needs, problems, and other phenomena with the passing years. This changing nature is particularly apparent when dealing with stress. The word seems to change meanings. In the preceding sections it has been shown that stress stems primarily from the mistakes, behaviors, and failure of persons other than the young victim of stress. That which can be done to reduce stress for infants and children must essentially be done by parents or caretakers.

Beginning with later childhood, individuals become increasingly responsible for the stress they experience. This is not a sudden about-face: the statement is "increasingly" responsible. Parents, teachers, and other persons are not suddenly relieved of responsibility as stress-producers. But increasingly, as youngsters gain independence of action and grow in cognitive ability to make decisions, they must accept some responsibility in producing or relieving stress.

Meditation and Relaxation

Because stress is a normal aspect of vigorous living, everyone should learn about its normality and how to view the stress they experience. Stress can be controlled, in part, by learning to relax. Meditation is often referred to as an antidote to stress. Before 1975 much of the information on the effects of transcendental meditation (TM) was anecdotal. Since then more than 200 colleges and universities have conducted experiments involving meditation. Some of the organizations, other than educational ones, that have explored TM are American Telephone and Telegraph, Connecticut General Corporation, Blue Cross Blue Shield, and the U.S. Army. Such studies have shown that measurable changes occur in heartbeat, blood pressure, lactate levels, and muscle tension. Goleman (1986) reports that meditation relieves inner tension, helps one deal with stress, lowers blood pressure, and improves physical and emotional health.

As a rule, the general picture seems to be that meditators enjoy enhanced psychological well-being, reduced anxiety, improved perception, and lower rates of drug and alcohol abuse. If we accept these results and bear in mind that growing up is often stressful, we should ask: What can students gain from learning to meditate in school? Schecter (1975) administered a measure of creativity to 80 high school students before and after half of them participated in a 14-week meditation program. He found that the meditating students made significantly greater gains in creativity than did their nonmeditating peers.

At a New York high school meditation was introduced cautiously. Trained meditators talked not only to teachers but to local Rotary, Kiwanis, and Lions Clubs, to parent groups and to the school board. Driscoll (1972) summarizes the results: "We believe that transcendental meditation has been of direct and positive help to students in our secondary school . . . Students, parents, and teachers report similar findings. Scholastic grades improve, relationships with family, teachers and peers are better, and, very significantly, drug abuse disappears or does not begin" (p. 237). Similar results have been reported with younger students, as early as kindergarten (Murphy & Donovan, 1983).

The implications for including meditation in school and college deserve more attention. If students can learn to alter their own physiological responses through meditation, they might prevent stress from interfering with their learning. This is a far better alternative than reliance on drugs or alcohol for coping with pressure. Moreover, behavior problems can be dealt with more constructively by encouraging misbehaving students to withdraw for a time, to relax, and find more healthful ways of coping with difficulty.

Transescence

The lack of uniformity which characterizes children as they move toward adolescence has been referred to by Eichhorn (1980) as transescence. He hopes by use of the word, to emphasize the wide variations between individuals in such things as emotions, intelligence, cognitive levels, and socialization, but particularly in the age when puberty occurs. He says, for example, "Transescents clearly cannot be classified as being at one or another stage of cognitive achievement, although some schools have consistently attempted to do so" (Eichhorn, 1980, p. 63). Parents sometimes do apparently recognize the phenomenon of transescence, although the recognition may be more a matter of personal convenience than an acknowledgement of individuality. For example, "You are too young to date," but later in the evening, "You are too old to neglect your responsibilities (school work or household chores)".

As children move into adolescence, the range of stress widens because of the increased degree of responsibility for self and others. Coping skills can be supplemented with rational approaches: verbal instruction, brainstorming, discussion, and problem-solving. For some people, during adolescence cognitive processes are emerging into reasoning and logical thinking.

ADOLESCENCE

Emphasis on Coping Skills

A list of potential sources of stress for adolescents suggests that their concerns pertain to life as a whole—not just a few specific problem areas. Stress may arise from:

> Racial prejudice
> Personal conflict with parents or peers
> Peer rejection or felt rejection
> Body image (pimples, changing clothes in front of peers)
> Dating (asking and being asked)
> Understanding schoolwork (willingness to seek help from teachers)
> Sexuality (getting involved or not)
> Keeping up with the crowd
> Using drugs and smoking
> Boredom and lack of purpose

Because there are multiple ways of dealing with any single source of stress, we shall place emphasis on the development of coping skills. This is not to suggest that adults (including parents, teachers, employers, advertisers, etc.) are no longer responsible for certain of the stresses adolescents encounter. We are saying that when persons reach adolescence the challenge of stress changes. Because peers are so important to them and because they can think reflectively, they must avoid the temptation to blame others. They should ask what they themselves can do. Discussion groups in high school, church, clubs, or wherever adolescents meet, provide access to verbalizing their problems, stating their views, and suggesting means of solving problems.

A few basic considerations for those who would conduct productive discussion groups are:

> Everyone must be given a chance to contribute even though not a fluent speaker or is one who does not boldly hazard risks.
>
> Situations in which young people have had actual experience and which have resulted in pleasant or unpleasant emotional reactions are better topics than are things which are highly general and over which they have little control.
>
> While no optimum group size has been determined, six to eight members is a workable number. Expert leaders can be effective with a dozen discussants; but as the number increases the chance of memberwide participation decreases.
>
> A metalinguistic emphasis is productive; i.e., talking about feelings, what others are thinking and drawing inferences are concerns that focus on the process of communication rather than content.
>
> Adult leaders must bear in mind that they are not the teachers, the court of last resort, or the summarizer. Their role is that of seeing that all get a chance to speak and to be heard. They are facilitators of growth.

Adolescent Wisdom

Because some things, in fact many things, in adolescent-adult transactions are in the form of roles and unilateral advice, adolescents sometimes become rebellious—on occasion—if not chronically.

Hence a potential source of adolescent stress is the trepidation parents and teachers have about young people making serious errors. And mistakes are made. Wise parents and teachers need to acknowledge that dependency should not, cannot, persist. When dependency is prolonged, adults might well admit their own contribution to failure to mature.

The authors recall a couple of sessions when it was a real temptation to interrupt with some advice. Near the end of a discussion on racial prejudice, when the temptation to give advice was avoided, one girl said, "My father bad-mouths blacks. He does not want them in the neighborhood, does not want to associate with them . . . I love my dad but I feel kind of sorry for him. He knows I have black friends but other than telling him I have such friends we can't discuss it. I definitely do not want to grow up with such limited views of people."

Another example of adolescent wisdom was provided in a discussion of drugs—marijuana, cocaine, and alcohol. Several teenagers described their drug trips and expressed their beliefs that the dangers were overstated. None of them seemed to have an answer when one boy said, "There are many things that I want to do academically, socially, and in sports. I have seen my buddies get dopey after temporary stimulation. Because I need all the intelligence, quickness, and strength I have to compete, I choose not to limit my natural power with drugs or alcohol." It would seem that this young man is on the way to being an achiever. He did, in fact, play varsity basketball and baseball and was on the honor roll in a high school enrolling some 2000 students.

Adolescent wisdom surfaces almost routinely in group discussions. The impact can be magnified by the group leader who turns attentively to such speakers and then emphasizes by requesting "Say that again," "Tell us more," or "Will you clarify that?"

Parents, teachers, and adolescents should disregard the longstanding myth that adolescence is necessarily a time of stress and strain, turbulence, and rebellion. Lipsitz (1980) emphasizes that the myth is itself productive of stress because it can instigate expectations of being overwhelmed. Stating that adolescence characteristically is not a stressful period does not mean that adolescents are without stress. Many dropouts, academic failures, delinquents, drug abusers, and potential suicides experience enormous stress. Typically these disturbed youngsters will not seek help but they are often responsive to it when they perceive it to be offered.

Boredom and Purpose

A major source of stress for many adolescents is their lack of a purpose or socially significant role. More than 50 years ago Benedict (1935) and Mead (1928) showed that in some societies there is no adolescence. One day these young persons were children and a few days later, after pubertal ceremonies, they emerge as adults performing work, marital, and parental roles.

More adolescents than ever before have part-time jobs and so experience a sense of purpose. Other teenagers find purpose in their school curriculum—the pursuit of excellence—or because they perceive those studies as being instrumental in their vocational or professional future. Still there are a large number of adolescents who claim parents should find something exciting for them to do. They seem to think that teachers should make school interesting; society should provide them with a purpose. These assumptions imply that stimulation must come from outside the self. Unfortunately, no one can force another to be interested and involved.

Boredom is more than just ennui and noninvolvement. It entails hazards. When we are bored, we may be inclined to indulge in risky ventures, promiscuous sex, drugs, and sometimes aggression. Perhaps all persons are endowed with varying degrees of a biological urge for sensation seeking. Boredom is in part a kind of sensory deprivation but it is also a frame of mind. We can talk ourselves into it. Some suggestions, applicable to young and old, for countering boredom are:

> Boredom is common but can be useful in prompting persons to identify a specific kind of richer life.
>
> Resist the idea of plugging into a variety of new pursuits.
>
> Instead, consider the source of your boredom and seek a specific activity, arduous if necessary, that has functional value.
>
> Consider the amount of risk you can afford and can tolerate (in changing goals, careers).
>
> Evaluate the role of fantasy in formulating new plans.
>
> Resist the temptation to withdraw into excessive television watching.

> Reach out to others, subjugate your self-concern to compassion and commitment to others.

EARLY ADULTHOOD

Adulthood: A Vaguely Defined Stage

Even though data are incomplete, much is known about childhood, adolescence, and old age. Comparatively little is known about that broadly conceptualized age-stage called adulthood. For the sake of convenience, we have used the ages from about 20 to 40 years to embrace early adulthood. The word "about" is focal because age-stages are functions of modal behavior, responsibilities assumed, and roles—rather than age alone. This is especially apropos in adulthood. Some people in this age bracket are still acting as children—returning home after a brief try at being adults, to fill the so-called empty nest that older parents were beginning to enjoy (Silden, 1982). Others not yet in this age bracket have assumed adult responsibilities since their teen years. There is little in our biological time clocks to identify adulthood (as there is in the case of childhood, adolescence, and to an extent in old age).

Multiple Challenges

Early adulthood is a stressful period because:

> Persons in our culture are on their own. They are through school or nearly so and are expected to be autonomous and responsible.
> They have the life-influencing decision to make of choosing a spouse.
> They must decide upon a career.
> They have to select a lifestyle.
> These crucial decisions must be made largely without help.

It is especially because crucial decisions must be made largely without help that we nominate early adulthood as the most stressful period of the lifespan. Adolescents had choices to make but they had available peer confidants, concerned parents, and professional counselor-teachers with whom to discuss their dilemmas. During early adulthood parents are no longer depended on because autonomy is part of the process of growing up. High school teachers are no longer readily available. Peer confidants are scattered by virtue of their having left school, gone to work, or attending colleges around the country. These conditions are products of culture—and people could avail themselves of counsel if they would. But they are generally alone and are supposed to be sane, strong, and self-sufficient. The situation is illustrated by a young lady who described her turmoil because of job and marital difficulties. She concluded her remarks with "Please don't tell me to see a psychiatrist. I'm not crazy."

One should consult a counselor with as much openness as one would consult a medical doctor about a chronic headache, upset stomach, or other physical ailment. We are beginning to see the wisdom of clinical help for such things as anxiety, stress, insomnia, and depression. Because 90 percent of circulatory and digestive ailments are alleged to be psychosomatic, psychologists and counselors ought to be consulted more readily for concerns that do not require physicians (Jenkins, 1985).

The Personal Role in Early Adult Stress

There is no sharp change in health and death rates for adults as there is in childhood (susceptibility to disease) or adolescence (high incidence of accidents and suicide). There is a slow rise from the mid-twenties through the forties in the incidence of heart disorders, circulatory problems, and psychosomatic illnesses. Authorities attribute this to choice of lifestyles rather than to the passage of years or job pressures. For example, the young adult suffering from hypertension is advised by his physician to pursue moderation, slow down, take it easy. Some may be frightened or wise enough to take the advice. Others are too tough or feel too much pressure to adopt a new lifestyle. Many would like to take the advice but simply do not know how to go about it. But the opportunity for choice is there. When persons do slow down and relax the ailment is usually brought under control.

There are a number of decisions of far-reaching consequences that must be made by the young adult. The major ones are marital pattern, parental roles and responsibilities, and vocation. All of these may produce stress. For example, the matter of choice and stress in fields of business has been referred to for years as "executive pressure." The young adult is required to sacrifice leisure, family, freely chosen social life to the demands of the struggle to stay on, or rise on, the corporate ladder. Some persons are able to do this; a few even seem to thrive on that lifestyle. However, most persons cannot ignore their spouse and children, neglect physical exercise, and restrict their social life to those places and people where they might make valuable contacts. Those who succeed seem to thrive on a competitive, all-out strategy, to achieve success. Others following the same competitive course develop heart dysfunction, high blood pressure, ulcers, and psychosomatic illnesses.

Occupational and Domestic Burnout

The term commonly used for succumbing to job pressure is burnout. This occurs in all occupations from laborers to professionals, and with about the same rate of incidence. The struggling mechanic is no more free from job pressure than is the young lawyer.

Most of the studies on burnout have focussed on the kinds of pressures that produce such symptoms as boredom, anger, dealing impersonally with others, indifferent performance and job dissatisfaction. It is estimated that half of all working people are unhappy with their careers (Hunt, 1983). The studies also deal with the physical symptoms of headaches, insomnia, digestive disturbances, and weariness. But the fact remains that some people in the same line of work are zestful, caring of others, optimistic, and cheerful. These satisfied workers may occasionally think of other kinds of work but they would not, if they were to do it over, choose another career.

Burnout should be seen in a broader context than just the workplace. The fact is that an increasing proportion of women are having to cope with the combined pressures of employment and domestic tasks. In effect many of them have two jobs. The average married woman with two children spends about twenty hours a week on the care of children plus another eighteen hours on household duties. When added to a full-time employment position this totals about eighty hours (Ornstein & Levine, 1982). Apart from the restriction this heavy commitment imposes on career development and productivity, it also results in a lack of leisure time.

Berk (1980) provides a view of how domestic tasks were allocated in the late 1970s. She surveyed 319 randomly selected women from intact families in the higher than average educated community of Evanston, Illinois. Nearly 90 percent of the 60 household tasks covered in their questionnaire were done primarily by the mother. Besides continuing to do what was traditionally called woman's work, and increasingly more of what men were once assigned (e.g., 60 percent emptied the garbage at least half the time and more than half went to the gas station as often as their husbands). These mothers report they usually do tasks one might expect of children. For example, 74 percent of them set the table regularly; 88 percent picked up toys as often as their children; and 70 percent participated equally in caring for pets. When mothers are assisted with domestic chores, what they receive is help; that is, the responsibility remains with the mother and the roles simply shift to a supervisory function. The nearly negligible child work around the house relates directly to the common parental complaint that boys and girls are not as responsible as kids used to be.

In a more recent study Nyquist, Slivkin, Spence, and Helmrich (1985) obtained data from 165 middle-class parents of first and second graders. Almost all of the husbands were employed as were two-thirds of the wives. Each participant completed a questionnaire which identified their proportionate responsibility for domestic tasks. Contrary to the assumption that there is a significant shift toward more egalitarian attitudes about role sharing in the home, this study showed everyday household tasks were divided along traditional gender lines. On the home tasks scale, in this study and for previous ones as well, scores were highly skewed with few women receiving any appreciable help from their spouse on a regular basis. It would seem that the most important sex role obstacle still be overcome is shared domestic tasks.

Choosing a Lifestyle

The need to choose a lifestyle is emphasized by the observation that in some ways early adulthood is truly a prime time in life. This period usually includes peak earnings because of energy available and seniority. Of course, one can still be unhappy if more affluent peers are picked for comparison. If comparison is made with those of earlier or later age-stages, there may be cause for gratification. Comfort (1981) also cites the matter of choice and responsibility for adults in quite a different context. Speaking of capitalizing on longevity potential he says that the overriding factor in prolonging life is ". . . not cellular but in the brain" (Comfort, 1981, p. 20). He refers to cumulative research which shows that psychological health (having social purpose, being active, and observance of common health rules—avoiding smoking, reducing tension, and avoiding excesses) is the basic factor in longevity. One can choose these.

The seriousness of adult choice in handling stress, adopting lifestyles, and developing values is emphasized by the fact that these years are ones of maximum cross-age influence. Both by precept and example, young adults greatly influence children, adolescents, and peers. Young adults are beginning to reverse the role of advisement with their parents. As adults learn to handle stress they simultaneously enhance the welfare of both the younger and older generations.

MIDDLE AGE

The Need for Self-Evaluation

From the time middle age became a specific concern of researchers and philosophers to the present, the necessity to examine roles and purposes has been emphasized. Lifespan scholars agree that a reassessment of personal direction is necessary in the middle years. They do not agree on just what goals middle-agers should pursue. We believe this lack of consensus is beneficial because it urges each individual to engage in processes of self-evaluation (Levinson, 1978).

At the same time a person must learn to cope with new perspectives about aging, it is important to take stock of old things, especially the longstanding illusions held about achieving occupational goals, about intimate relationships, about what one really wants out of life. By age forty persons know about where they are and how much further they can expect to go. If there is failure, there is a certain regret and adjustment that goes with accepting that knowledge. But even those who have avoided failure, persons whose occupational goals have been met and surpassed—they too report having to cope with feelings of futility and meaninglessness. Indeed discontent is so common an experience among middle-agers that researchers regard its occurrence as a normal part of personality development (Moos, 1985).

What seems to happen at this point is that certain of the previously obvious truths, the basic beliefs and supporting values suddenly come into question. The sense of disparity between "what I have reached and what I really wanted" instigates the soul searching for "what I want now." It is obvious that most people find it hard to amend their dreams. Many hold on to fantasies that should be replaced; uncertainty prevents others from making any change at all. Yet the necessity to activate a more appropriate dream becomes the major developmental task of middle age. The need is to restructure earlier aspirations in such a way that the worthwhile and mature dimensions of their original form are retained while new priorities consistent with being middle-aged are added.

Discontent provides the motivation for changing the dream but not the direction for change. Because people differ in the source of discontent, they vary in the kinds of questions which are used for self-evaluation. Where marriage is the focus, one may begin to consider what is missing in the husband-and-wife relationship. Frequently persons of this experience ask themselves: Do I want to stay with my spouse now that the children are grown? Should I have an extra-marital affair? Of course, certain people express no dissatisfaction with the family. Instead what disappoints them most is the situation at work. Whether they are bored or pressure-ridden, their questions are similar: Do I want to go on doing the same kind of thing for the years that remain? Is it too late for me to change jobs? Too late to start a different career? Industrial sociologists are finding that alienation from the jobs has become common and that such behavior is destined to increase with the monotonous tasks being created by automation.

The physical changes of middle age also cause discontent. Declining efficiency of the body leads some to become concerned about heart disease, others to consider a more moderate lifestyle, and still another group to be preoccupied with changes in appearance and its effect on interpersonal relations. Here the questions are: What habits must I discontinue or renew to improve my chances for good health? How can I keep my body youthful and attractive? That these concerns are widespread is shown by the army of middle age persons who frequent health clubs, exercise programs, diet clinics, and cosmetic services. One additional source of discontent deserves mention. In this case the issues for reflection involve the discrepancy between a person's religious or philosophical values and less mature actual behavior.

It is relevant to note that people differ not only in the source of discontent but its intensity as well. In our view the severity of discontent accounts for why the practice of self-evaluation during middle age brings only slight changes in some persons and yet significantly alters the future for others. Because most of the writing about middle age has been done by clinicians whose contact with healthy people is infrequent, it has been customary to report the middle-age experience almost exclusively in terms of crisis. By this version, the reflective self-evaluation of persons seldom leads them to a better life but instead introduces long-term depression, alcoholism, sex disorder, divorce, obesity, and a host of other undesirable conditions. Admittedly these circumstances match the lives of some people. However, we agree with Neugarten (1980) who contends that middle age can be a time for personality development which initiates maturity. This transition is reflected by a higher level of intimacy in marriage, increased job satisfaction, a deeper need for relationships with children, or stronger commitment to moral principles. This optimistic view is supported by Nicholson's (1980) survey of middle-aged persons who were asked to identify the years since adolescence they found to be most stressful. None of the forty- to forty-five-year-old respondents described their present age in these terms.

Focus on Personal Development

Some problems of change are rooted in culture and the times—they are not universal or timeless. For example, involvement in parenting may not be so satisfying to middle-aged women now as it was in earlier times when longevity was shorter. Women were "old" at age forty-five in the beginning of the century. Today women of forty-five still have many years to consider after childrearing and are increasingly involved in the business world. To an extent the longevity factor is also pertinent to men, even though their identity continues to depend mostly on their job. Both men and women must find alternate purposes to which they can devote themselves in later life. Bardwick suggests these obstacles deserve consideration:

> Americans resist acknowledging the inevitability of their aging.
> Adapting to age-stage changes requires reevaluation of goals and values.
> Changing jobs and spouses tends to evade the basic requisite, inner change of accepting middle age.
> Those who accept community responsibilities are

"... more realistic, adaptive, integrated, and complex ... Some will be wise rather than merely smart, humane instead of egocentric, the mentor rather than the student" (Bardwick, 1978, pp. 134, 136).

The last item in Bardwick's assessment of middle-age stress and its resolution seems to be another empathic statement of the need to switch to the goal of generativity, from self-concern to concern for others. Again, we must beware of the all-inclusive generalization. There are many people for whom psychological, physical, and social health are maintained by the continued pursuit of individual or occupational goals. Most persons, however, will find a more sure and safe avenue to health in the pursuit of generativity.

LATER ADULTHOOD

Sources of Stress

Much of the literature about stress during later adulthood emphasizes the reluctance of some people to accept retirement. Prior to 1960 it was necessary for most people to work long hours and a full lifetime just to survive. No precedents were set for the constructive use of massive leisure time. Today shorter work days, more vacation, and less lengthy work lives may still result in quite adequate life income. In addition, the increasing number of older persons urges retirement to make room for younger workers. We are referring to paid work. There is still much for older persons to do in volunteer work for both their age mates and in performing service for the younger, disadvantaged, and disabled population.

The belief that retirement inevitably leads to a loss of identity and thereby contributes to ill health is unconfirmed. Despite common anecdotes about older persons who lose their health shortly after retirement, there is no research data to support this view. Neither compulsory nor voluntary retirement seem to have a negative impact on health (Brubaker, 1985).

A potential threat which stresses many older people is the increased probability of losing one's mate. There is help available for dealing with grief, coping with the feelings of loneliness, anger, and depression that result from the death of a loved one. Alcohol, drugs, and withdrawal are more likely to compound the problem than to dim its existence. The traditional approach to grief has been to resort to prayer and religious assurance. These resources can be augmented by participation in self-help groups or counseling that guide the bereaved into new patterns of human interaction.

Another potentially stressful factor is change of residence. Some concern has been expressed for retired people who move from a large to a smaller house or apartment, from a now not-so-familiar neighborhood (friends gone and strangers coming in) to a retirement community. When counseling is available the stress of moving can be reduced. Group discussions are especially helpful because of the support of those who have "gone through it." The anger, sadness, anxiety, and sense of loss that may accompany moving must be faced, discussed, and accepted. Cry a little, laugh a little, and then settle down to making friends and developing a new and appropriate living pattern (Dentzer, 1985).

Gober and Zonn (1983) report that the stereotyped view of elderly migration is unsubstantiated. They studied newcomers to Sun City, Arizona, one of the nation's largest communities designed exclusively for older persons. The findings revealed that elderly people do not move impulsively. In fact, some move to get away from the stress of becoming too much caught up in the lives and problems of their adult children. Most migrate because (1) they have a close relative in the place moved to, (2) they have friends there, (3) they like the amenities available (recreation, medical facilities, shopping ease), (4) they enjoy the climate, and (5) a trial period proved satisfactory.

There is misunderstanding about stress and the financial circumstance of older people. The majority are economically capable of taking care of themselves. Those who are poor in early and middle adulthood are the ones who provide the base for the myth that aging and poverty necessarily go together. Linden (1985) points out that:

> The per capita income for 50–65 year olds is 20 percent higher than the national average.
>
> Americans over 50 years of age own a commanding 77 percent of all financial assets belonging to households.
>
> The average savings of 65–70 year olds is double the national average.

Given these indicators of affluence, why is it that older adults express so much concern about money? Usually the worry stems from an awareness of increased personal vulnerability to illness and the potential health care costs for recovery. The dread of depleting life savings is compounded by a fear of becoming a burden to one's adult children.

A growing source of stress for older adults is the unhappiness of their children, especially when divorce occurs (Spanier & Hanson, 1982). Depending on their affection for the son-in-law or daughter-in-law and their estimate of the marriage benefits, older adults may feel an acute sense of loss or relief. For those whose daughter is assigned custody, caring for the grandchildren until the situation settles down is one way to be helpful. Single parents often suffer from inadequate income while grandparents are more affluent than other age groups. When grandparents provide a gift of money to single parents, they can arrange leisure time and a chance to do something special with their children.

When divorce severs the tie between a husband and wife, it can also separate their children from one set of grandparents. Some custodial parents may want to be rid of the spouse and that person's relatives as well. Fortunately, the law allows grandparents the right to petition the court for visitation rights when there is a divorce. It is recognized that grandparents can provide children with a sense of continuity and dependable affection. Visitation is approved if the court feels that the grandparent request does not undermine the mental health of grandchildren (Kornhaber, 1986).

Perspectives on Stress

Stress, as it has been emphasized, depends greatly on personal viewpoints. Gerontology continuously adds more hope for reduction of stress as harmful obstacles are displaced. For instance, assumptions about senility, physical deterioration, biological involution and social ostracism have been proved incorrect and damaging. Actually there is more difference among persons in late adulthood than there is between them as a group and the middle aged. There are some senior citizens whose health is better than younger persons. They have visual and hearing losses but these can be compensated for by glasses and hearing aids.

One way for older people to counteract these harmful myths which promote stress is to make the first move toward intergenerational contact and communication. By doing so they can modify the public view to the advantage of everyone. For example, in Chicago, senior citizens serve as resource persons for an innovative social studies program intended to rectify the imbalance of time students usually spend on studying the past compared to the future. The curriculum entitled "Growing up and Growing Older" includes aging as one of the most predictable events in the future of today's children. As the twelve to fourteen-year-olds discuss their concerns and questions about the stress and satisfactions of adult life, the senior adults are often called upon to react and share their experience.

Chapter 8

ACQUIRING SOCIAL COMPETENCE

"No man is an island, entire of itself; every man is a piece of the continent, a part of the main ..." These often-quoted words of John Donne express the vital importance of socialization. Persons cannot live by themselves. Nor, it must be added, can babies live without the prolonged care of mothers. One of the delightful recent discoveries about human development is the high degree of infant social competence. If you doubt it, reflect on how easily a baby "steals the scene" from parents, clergy, Hollywood stars, and celebrities.

The fact that infants are instant socializers does not mean that those same skills are retained and are effective in subsequent years. Witness the large number of social failures: selfish children, alienated adolescents, martial dissolution, criminals fighting public restraints, and persons of all ages who are so at odds with others that they withdraw—physically or psychologically—from their social milieu.

We are not sure that social failure is any greater today than it has been in the past. Statistics on delinquency, divorce, and alienation suggest that the rate of failure is climbing. To the extent that there is increasing turmoil and estrangement, we postulate that a key factor is the amount of attention given to personal identity. It seems identity is a paradoxical phenomenon. On the one hand, a healthy identity is basic to competent socialization. But an overemphasis on self can undermine personal development.

The purpose of this chapter is to (1) learn how self-esteem and group welfare are related; (2) examine some of the processes by which people learn to accept and get along with others; and (3) consider the need for continual adjusting to new roles with the passage of time.

Socialization in Perspective

A Concept of Socialization

Socialization refers to the acquisition of age-appropriate behaviors, values, and attitudes that are expected of persons who are accepted and esteemed in given social milieus. Parents, peers, teachers, and other key individuals are continuously involved in modeling, teaching, and conditioning (by means of approving or disapproving) growing persons to enact and internalize the expected behaviors or personality traits. The major target of studying human development is to produce, or become a socialized person. Obviously some fail. Those who withdraw are called asocial persons. Some who fail attack others and seeking to avenge their failure develop antisocial behavior and motivation.

Neugarten and Hagestad (1976) postulate—as did the biblical Psalmist—that there is a timetable for ordering life events. There is a time for intensive learning, a time to marry, to raise children, to be a contributing member of society, and time to retire. The concept of developmental tasks at various age stages relate mainly to steps in socialization processes. Socialization is acquired not innate. There are two major theories involved. *Social learning* refers to those traits acquired through modeling, conditioning, training, and instruction. *Social cognition* refers to those aspects of socialization dependent on understanding, interpreting, and evaluating other persons' feelings, goals and probable actions. This aspect of socialization is dependent on, or parallel to, cognitive development (Maccoby, 1980).

INFANCY

Social Awareness

It was believed, early in this century, that infants were overwhelmed by being born into a big, buzzing world of confusion. Studies in the last decade tend to contradict this belief. Infants, in fact, seem to be quite aware of a number of things and have distinct preferences. While newborns have almost no vision (about 20/500, or a "legally blind" rating according to initial assessment) their sense of sight develops rapidly with use. By eight weeks they can differentiate between shapes and colors. At three months stereotypic vision starts to develop. Babies prefer the complex over the simple. Given a choice, they will look at a checkerboard surface rather than a plain one, or a bull's-eye target in preference to stripes. Infants arrive with a wide variety of auditory reactions. Unlike eyes, ears begin functioning long before birth. Newborns are comforted by and fall asleep faster to the sound of a human heart beat. They prefer female voices and recognize the sound of their mother's voice almost immediately. At one month of age they can differentiate between the sounds in virtually any language. Babies have a sophisticated ability to put sounds into categories. For instance, they know which sounds communicate and which do not (Friedrich, 1985; White, 1985).

Infants have a number of discrete emotions that suggest a surprising degree of social awareness. Newborns show evidence of interest, startle response, distress, and disgust. Within four to six weeks they can give a distinct social smile. Before they are a year old they show anger, surprise, sadness, fear, shyness, and contempt. These appear to be confirmed by photographs, cross-cultural studies, and consensus of trained and untrained observers. It is hoped that mothers and other caretakers can be taught the distinctiveness of these various emotions and social inclinations; and then become more sensitive to, and specifically responsive to the moods and needs of infants (Brazelton, 1983; Trotter, 1983).

Another cause for respect of the infant as instant socializer is the phenomenon of imitation. During the first weeks of life infants can imitate an adult in opening the mouth, widening the eyes, and sticking out the tongue. Bower (1982) says these acts are unique to human infants and are not confusing to them. They do not open their hands when they see adults open their mouths. They do not widen their eyes when they see adults stick out their tongues. They imitate. Imitation "...surely indicates that the infant is aware that it is human. By imitating, the infant is showing us that it knows it has eyes, mouth, tongue, hands and that these parts of itself correspond to the same parts of us. I would argue that imitation is an affirmation of identity, evidence that, at some (however primitive) level the infant knows it is one of us" (Bower, 1982, p. 259). Infants can imitate even where there is a delay between stimulus and response. As described in Chapter 5, when an adult sticks out his tongue and the infant cannot do so because of a pacifier in its mouth, the tongue will be stuck out after the pacifier has been removed or pushed out.

Attachment and Trust

The basic phenomenon in socialization during infancy is attachment. This refers to the infant's strong preference for, and dependence on, another person for physical closeness—a feeling of being protected and recognized by a specific person. Newborns briefly are unattached but as they become more cognitively mature they fix on a specific person; given the chance, they cling to and depend on that person. If these bonds of affection, dependency, proximity, and constancy are not formed between the age of six months and eighteen or twenty months, the chances for normal human social development are seriously damaged. In short this is a critical period of infant development. Bowlby (1980), who has been studying and reporting on mother-child interaction for three decades regards attachment as one pole of human experience—loss is the other pole. The importance of attachment was stated some time ago in unequivocal terms by Fraiberg (1967, p.57): "If we take the evidence seriously we must look upon a baby deprived of human partners as a baby in deadly peril. This is a baby who is being robbed of his humanity." She paints this word-picture of horror because of the frequency with which violent criminals, asocial and antisocial persons experienced little or no warm parenting contacts during the critical months for forming human bonds. They often become hollow people, persons who have no human feelings, or a need for others. They are virtually unresponsive to their fellows.

Because of the lifelong significance of attachment as a basis for socialization some description of its formation is desirable. The familiar term "tender loving care" is more than a cliche. It is as much needed as food itself for normal development. TLC is revealed by all mammals, normally. However, because of a rising demand for surrogate day care of infants, the need for continuity of persons who touch, bathe, feed, and dress the baby must be emphasized. The same persons should hug, squeeze, hold, and talk to a baby, repeatedly and habitually. Fortunately, engaging in these activities makes the caretaking person more loving. It is almost as if loving follows rather than precedes the caretaking (Brewer & Brewer, 1985).

The high incidence of infant neglect, abuse, and abandonment shows that devoted motherhood or fatherhood is not something with which persons are congenitally endowed, just because they are physically mature enough to procreate. Some young adults do have a healthy eagerness to become "good" parents. Others have to learn it by deliberate intent and by devoting time to the process of an infant's socialization. Quite possibly, time is more important than love because children can be loved without being given time. Moreover, the time given to caring, caressing, and talking to the infant may lead to love. Going through the motions may lead to the emotion. Unwed mothers, for example, have on occasion, prior to seeing and holding their newborn decided to give it up for adoption. After seeing, holding, and speaking to it they have decided that they will find some way to retain that mother-child relationship—to exercise the human bond that has so quickly been initiated (Klaus & Kennell, 1983).

Early Childhood

Territoriality and Early Socialization

Peer group influence begins at an earlier age than ever before for most children. Because of new work opportunities for women, the desire for greater family income, and the rise in single-parent families, more boys and girls are being thrust into group care settings well before they prefer it. Half of all employed mothers return to work within a year after childbirth. The net results, in most cases, is that children are spending greater time with peers than any previous generation; they are age-segregated as never before. Although this condition is better than being left alone, it also presents the possibility that immature models may overshadow adult guidance (Bronfenbrenner, 1984). This concern gives rise to an important question: How can those who care for young children in group situations help them to have a beneficial effect on one another?

First, let's examine the nature of interaction among preschoolers. Parents and teachers frequently express concern about the selfishness and possessiveness of young children. It is common practice to urge them to share and cooperate. This custom ignores a pertinent phenomenon known as territoriality, the inclination for creatures to declare, usually nonverbally, a certain space as their own. The behavior can be observed throughout the animal world. Coyotes and wolves mark their territory by leaving their scent on the boundaries of their space. Fish will attack larger fish who invade their space. Cats will do the same. A similar intention to possess space can be observed among human beings. People stake out their land with fences or survey posts. We become irritated when a drive cuts into the 30-foot space we have allowed between us and the car ahead. We dislike and draw away from the person who gets closer than the 18 inches allowed in face-to-face contact. Exceptions are our spouses and our children (Ardrey, 1966).

Dominion Play and Peer Relationships

When a child asserts a claim over a plaything or play-space, it is known as dominion play. This kind of territorial play is normal for children between two and six years of age. Nicolayson (1981), a kindergarten teacher, regards children's need for privacy as a factor in the development of self-concept. She advises that teachers acknowledge the fact of territoriality and validity of dominion play. When dominion play gets in the way of a smoothly functioning day-care or preschool group, the teacher and children should engage in a dialogue about mutual rights, the key to early socialization. Consider the example of Carol and Dale, both four-year-olds enrolled at a day-care center. Carol, in tears, approaches the teacher to report that Dale will not let her play with him. After acknowledging that she understands Carol's feelings, the teacher suggests that they talk with Dale. After Dale explains that is making a zoo and doesn't want any helpers, the teacher turns to Carol who indicates she wants to be his partner anyway. Because Dale is not infringing on anyone else's territory, the teacher defends his right to privacy by telling Carol that she will have to play with someone else or by herself. To force Dale to admit Carol against his will would violate his right to privacy and lead to inauthentic relationships between the children. In similar circumstances the teacher would defend Carol's right to privacy. When boys and girls cannot look to adults to defend their privacy, they develop a sense of helplessness rather than self-confidence.

Conversely, there are times when a child's dominion play interferes with the rights of others. Then limits must be set, not to deny a youngster space but to restrict it such that others can also satisfy their needs. Jim, age four, visited a roundhouse at the railroad yard over the weekend. On Monday morning he decides to build a replica out to blocks. Unfortunately, his project is placed so close to the block shelves that nobody else can get their play materials. After observing the unsuccessful attempts of Jim's peers to have the roundhouse moved, the teacher decides to approach him. "Jim, the reason everyone wants you to move your roundhouse is because they can't get to the block shelves." Jim points out they better not touch it or the roundhouse will fall down. The teacher makes another suggestion: "Jim, can you see any place to move your roundhouse so the other children will be able to play too?" Jim is adamant about not moving his structure. Then the teacher says "I know you don't want to move it but we'll have to find another spot." Undaunted, Jim states that "It's already built and so it can't be moved." The teacher asks if Jim wants to find the new location for the roundhouse or if he would like her to identify a good one. Again, no deal. Next the teacher proposes "I'll help you move it over by the window or you can do it yourself." Jim replies, "No." Without further comment the teacher dismantles the roundhouse and moves it toward the window so that everyone can gain access to the blocks.

Most children will more readily accept suggestions than Jim but even when they do not, providing them face-saving alternatives is a better method of teaching than punishment, embarrassment, or specific commands. Boys and girls can learn to work alongside (not in) the private space of peers. This accommodation of mutual rights is essential for social competence. And, as a rule, when a child's right to privacy is respected, he becomes less defensive and may soon welcome play with the same children he recently rejected.

As children get older they may decide it is all right to play with one companion but no more. By excluding all others the players make it known that "Two's company, three's a crowd." Teachers and parents often disapprove of this behavior. Instead of expressing friendship preferences, it is felt that children should learn to like everyone. Obviously this expectation is unrealistic and seldom met by adults themselves. A better responses is to accept the fact that, in most circumstances, children should be permitted to choose their own friends. Because friendship requires a certain amount of privacy to develop, it is appropriate to honor the preferences of youngsters to be alone together.

Guidelines for Dominion Play

Parents and teachers who consider the following guidelines can more readily support social development during the preschool years.

- Little by little, the young child gains a senses of mastery. You can help by respecting the need for privacy, ownership, and control of space.
- Encourage boys and girls to respect the privacy of others. Set limits for mutual rights.
- As the child grows older, he may choose to welcome another person into the play situation. Whenever possible, allow boys and girls to make the decision about who enters and who does not.
- Help both children in the interaction to express their feelings in an acceptable way to each other.
- Conflict learning begins with early socialization. But as a rule adults choose to approach child conflict as judges whose function it is to reach decisions about guilt and punishment. This strategy denies children an opportunity to make their own decisions and it fails to introduce the necessary alternatives they need for making good choices. Remember that an important part of decision making which children must learn is how to generate alternatives (instead of hitting people or violating their rights).
- So, before rushing in to solve a conflict between children, take a few moments to observe the dynamic and find out what is really happening. Encourage children to solve their own conflicts.
- If intervention is necessary, generate a range of face-saving alternatives that restore mutual rights. This task will be difficult in the beginning because it re-

quires creativity on your part but, with practice, you can be the model that children need.

Social Incompetence and Misbehavior

What will happen to the 11 million children enrolled in day-care and preschool settings if their caretakers do not understand dominion play? What will happen if they overlook the importance of mutual rights, deny privacy, force sharing, and resolve child conflicts by coercive means instead of raising questions and generating face-saving alternatives? The evidence suggests that socialization will be adversely affected. In North Carolina, children who started in day-care at the age of six weeks and continued until age five were compared with a control group who had been cared for at home. When the children were measured for aggression, day-care population was found to be 15 times as aggressive as the controls. This was not a matter of greater assertiveness to stand up for personal rights but rather an inclination to verbally and physically attack others. Those who had been in day care since infancy were also found to be more easily frustrated, less cooperative, more egocentric, less task-oriented and more distractible. Farran (1982) concludes that the emerging picture is of children who have not developed self-control and are ready to fight to resolve difficulties.

Furthermore, the dangers of not learning mutual rights during preschool seem to be international in scope. There have been eight major investigations involving the impact of day care on social development in the United States, England and Sweden. Seven are consistent in finding that day-care children are less competent in socialization skills than children who do not attend (Tizard & Hughes, 1985). Later, in elementary school the consequences of social incompetence include invading other children's space, going up and down the aisles bothering classmates, distracting peers, taking things from others, and preventing the classroom conditions necessary for optimal learning.

MIDDLE AND LATER CHILDHOOD

Belonging and Social Prejudice

Helping children to see themselves as successful is one way to support their personality development. Children also grow by learning to see other people in positive ways. During the preschool years social prejudice is at a minimum. The activities of young children are usually noncompetitive and consequently do not threaten self-concepts. Because norms for intellectual performance have not yet been established, advantaged students do not feel superior; nor do those who are below average feel inferior. Throughout the primary grades, parents are the main source of a child's self-esteem. Youngsters accept as their own the value they perceive their parents to have attained.

Things change when boys and girls enter the intermediate grades. Now they can no longer depend on what their parents have accomplished as the basis of their pride in themselves. Instead, from this time onwards, children must earn their own social standing. In school they begin to encounter competition and failure, as well as classmates' critical and sometimes unfriendly remarks. Peers are quick to seize upon every weakness, their comments ranging from direct insults like "Fatso," "Skinny," "Retard," and "Four-eyes" to more individual and subtle but equally critical comments. As a result, intermediate grade children often display both a sharpening sense of self and a decline in self-esteem. One important aim of elementary school programs should be to provide experiences that enable children to see themselves favorably without having to find fault with others.

As peers replace parents in supplying norms, roles, and models with which to identify, boys and girls naturally look more to agemates for acceptance, information, and emotional support. Because it is so important to belong, children seldom question or challenge the ways of their group; instead, they strive to conform. As members of a group, they feel reassured if they behave like all the others without quite knowing why. This tendency to adopt values and attitudes without knowing the reason can lead to acceptance of viewpoints that children would otherwise recognize as wrong.

Do you recall the adage "Sticks and stones will break my bones, but names will never hurt me?" Most people have repeated this saying or some variation of it, but few of us are psychologically equipped to endure insult. Names do hurt and the hurt can be incapacitating. Normally we may seek retractions from individuals who insult us, or we may try to ignore them. But what if the name-callers include several people in our peer group? Then being called weird or sick does hurt, sufficiently so that most of us are inclined to alter our behavior in the direction of compulsive conformity.

For a group to have special significance, its membership must be somewhat exclusive. Beginning in about the fourth grade, children with similar backgrounds and standards tend to stay together, refusing to accept others who are different from themselves into their group. This clique behavior often leads to intolerance and prejudice. Indeed social prejudice is more common among ten-year-olds than among high school seniors.

Although it is gratifying to know that social prejudice is likely to decline as education increases, the effects of prejudice during the intermediate grades can last a lifetime. Those who serve as targets must divert some of their energy from the healthy development of their egos to self-protection. Prejudice and self-concept are inseparable; the way others treat us tell us who we are. To suggest that minority children or other victims of prejudice disregard the low estimate members of the dominant group have of them is to presume that children can avoid being affected by peers. The fact is how others feel about us influences how they treat us and consequently how we feel about ourselves (Elkin & Handel, 1984).

Accepting the Handicapped

In 1975 Congress passed the All Handicapped Children Act. What makes this legislation remarkable is its duration; it is intended to govern the nation's educational policies in perpetuity. The law was needed because more than a million children were being excluded from public schools and appropriate educational services were being withheld from more than half of the 8 million with handicaps (defined as mentally retarded, hard of hearing, deaf, speech impaired, visually handicapped, seriously emotionally disturbed, orthopedically health impaired or suffering specific learning disabilities). Some of the significant elements of the law designed to assist this 10–12 percent of the school population are:

> A free public education is made available to all handicapped children between the ages of 3 and 21.
>
> Access to education within a regular classroom is guaranteed unless the person's handicap is such that services cannot be properly offered there.
>
> School placement and other educational decisions will be made only after consultation with the child's parents.

Parents can examine and challenge all relevant records bearing on the identification of their child as handicapped and on the kind of educational setting in which the student is placed.

Previous federal legislation aimed at the elimination of architectural barriers to the physically handicapped will be applied to the funding of school construction and modification.

The integration of exceptional children into regular classrooms is known as mainstreaming. This approach allows handicapped children, including those who are mildly retarded, to remain with normal classmates through much of the school day. About 3 percent of the entire school age population is retarded, but nearly 90 percent of these persons fit the mild classification, based on IQ scores within the 50–70 range. Unlike the other mentally retarded individuals whose condition is referred to as severe (0–29 IQ) or moderate (30–49 IQ), the mildly retarded can, with help, be expected to hold a job, marry, and function responsibly in the general society. When mainstreamed, they leave the class only for certain subjects that are taught by special educators. This arrangement avoids isolation and recognizes that we cannot prepare children for life in a culture of diversity if the atypical are segregated in the schools.

For mainstreaming to work, social learning must take on a new significance for everyone. Unless social development in the education of normal children is given the same high priority presently reserved for cognitive development, exceptional children are likely to suffer rejection in the classroom. When schools insist upon uniform expectations for achievement, the emphasis is on teaching the retarded and other handicapped students to become more like everyone else. If, however, the goals of special education are broadened to include educating nonhandicapped youngsters to accept and honor differences, the focus changes. Students can learn to value people independently of their cognitive achievement or physical condition. The transition of this kind of learning, which essentially concerns personality development, will not be easy for an institution that has built its reputation primarily by supporting cognitive development, but it must be made.

Most of us will eventually encounter some handicaps. It can be predicted that a certain proportion of normal children will acquire impairments or disabilities, whether through accidents, illness, increasing age, or changing social conditions. Furthermore, some non-handicapped children will grow up to become parents of handicapped boys and girls or employers of impaired workers. These are ample reasons why learning to accept differences should become an identifiable element of the school curriculum. The innovation is essential if individually and collectively we are to convey a sense of worthwhileness to those who are unlike ourselves.

Values Clarification

Becoming socialized requires the development of values. Children growing up today derive their values from many sources, including parents, peers, television, heroes in music, sports and movies, the school, church, and relatives. Sometimes these sources are in conflict and their credibility is called into question. To further complicate matters it has become difficult to obtain consensus about what constitutes deviance in a pluralistic society. If virtually all behavior is to be judged acceptable, then nothing is immoral and aberrant conduct can be described as just different. Another factor promoting confusion is that during the past decade our changing moral climate has been paralleled by a restructuring of the legal system, from one that focused on fault and personal responsibility to a no-fault system that often times separates fault from obligation. There is no-fault insurance, no-fault divorce, and no-fault bankruptcy. In criminal cases, offenders may avoid responsibility when lawyers show that incriminating evidence was gained by illegal means. Even murderers can assert they are not responsible for their crime because it allegedly occurred during a period of temporary insanity.

Given the increasing ambiguity about right and wrong, fault and responsibility, it is not surprising that some youngsters are bewildered, even overwhelmed, by the kinds of moral decisions they must reach. It is unreasonable to suggest that until boys and girls are older and more mature, their parents or surrogate will make choices for them. The fact is that deciding whether to get involved with drugs, intimate relations and crime is occurring at an ever earlier age. Fortunately, when ten- to twelve-year-olds are encouraged to develop a coherent set of values, determine personal goals and set a course of direction, they are less likely to be misled by those who would otherwise take advantage of their vulnerability, sense of confusion, and indecision.

According to Simon, Howe, and Kirschenbaum (1978) none of us has the "right" set of values to pass on to other people's children. But by providing children access to an approach known as values clarification we can help them learn how to examine their own feelings and beliefs so that the decisions they make are conscious, deliberate, and consistent. Teachers who offer values-clarification techniques must abandon the traditional reliance on lecturing, reward, punishment, and other indoctrination strategies for imposing values. In order to motivate all students it is essential to emphasize their personal growth as well as intellectual achievement. Somehow schools must provide students with the skills they need to sort out the conflicts and confusion in their lives and enable them to identify the things they value. Until these goals are accomplished, some teachers are destined to continue describing many of their students as "difficult to work with and unmotivated."

Students who participate in values clarification are asked to respond to worksheets and questionnaires which encourage them to consider alternative ways of thinking and behaving in response to common problems of personal choice. The pros and cons as well as consequences for each alternative are carefully weighed. The teacher's task is to listen, reflect, attempt to elicit responses from all the discussants, call for repetition and clarification of their statements, and see that no one belittles or makes fun of another person's point of view. Finally, the teacher provides options to demonstrate values, inside the classroom and out of school. Only as students make their own decisions and assess the outcomes do they truly develop a viable set of values.

The seven basic processes which comprise values clarification require students to:

1. Choose their values freely;
2. Choose their values from alternatives;
3. Choose their values after weighing the consequences of each alternative;
4. Prize and nurture their values;
5. Share and publicly affirm their values;
6. Act upon their values;
7. Act upon their values repeatedly and consistently.

There is considerable evidence that students who engage in values clarification become less apathetic and less conforming; more energetic and self critical; less indecisive and more likely to carry out personal decisions. For some low achievers, the results also include greater academic success as they come to develop an internal locus of control.

Latchkey Children

When mothers are employed their problem of arranging supervision for children does not end with the beginning of elementary school. Most parents go to work well before the time sons and daughters leave for class. Later, in mid-afternoon, the students are dismissed and return home several hours before their parents. Long and Long (1983) of Catholic University first coined the term latchkey children to describe the phenomenon of the 1980s. Currently there are 10 million of them and the number is expected to grow as single-parent households increase and more mothers enter the labor market. Two-thirds of all mothers with elementary children are now employed or seeking employment; that figure is projected to be three-fourths of all mothers with elementary children by 1990 (Nolan, 1984).

Regardless of their income, parents are concerned about the possible negative effects that may accompany the latchkey experience. Although many families seem to do all right, the overall evidence is unfavorable. In terms of risk the unsupervised child (1) may feel bad (lonely, rejected, nervous, afraid); (2) may act badly (vandalize, shoplift, or abuse peers); (3) may develop poorly (academic problems, personality difficulties); or (4) may be treated poorly (suffer accidents or abuse by peers or adults). These anxieties about latchkey children are well founded for it has been determined that they do contribute beyond their proportion to the numbers of students who have problems with reading, drug abuse, absenteeism, dropping out of school, and delinquent behavior. A report by the California state senate's Office of Research indicates that most fire deaths among children occur when children are left alone (Packard, 1983).

For mothers who have the option of working part time rather than full time, there is evidence that being away from home less might be wise even when children reach adolescence. Whereas younger latchkey children complain of fear, loneliness and boredom, a study by Long and Long (1983) of 400 twelve-to-sixteen-year-olds in Washington, D.C. found 40 percent of the teenagers living in single-parent families had been involved in heavy petting or sexual intercourse at home after school. Fortunately there are a number of independent but mutually supportive efforts underway to assist latchkey children and their parents. These initiatives include:

1. Employer child care. Most of the employer involvement stems from in-house surveys showing that working parents consider infant and after-school care to be their biggest need. Over a thousand companies provide some form of support. Sometimes the employer administers a childcare center at the work site. Researchers from the University of Minnesota were hired to determine the employer benefits of this approach. There was less absenteeism, less tardiness, easier recruitment of new employees, higher morale and increased productivity. The latter result is commonly mentioned because mothers who lack proper supervision for their children often spend a considerable amount of company time after 3 pm on the phone checking on their home situation.

2. Survival Training. The American Home Economics Association is promoting short-term courses in elementary school that teach boys and girls survival skills. Such things as food preparation, safe methods of cooking, how to shop for nutritious foods, doing the laundry, and keeping the home secure from intruders are considered. Similarly, the Boy Scouts of America has designed a program to assist 6–12-year-olds cope with being home alone. The focus concerns preparing snacks, taking care of younger brothers and sisters, dealing with emergencies, home safety rules, and how to get help when it is needed. These and other programs also provide parents with information regarding reasonable expectations for children who are left alone at various ages, the kinds of rules that should be established, and ways to encourage self reliance without overwhelming boys and girls with too much responsibility.

3. The Friendly Listeners. More than 200 telephone reassurance programs are available across the country to give support to lonely, fearful, or bored children who are home alone after school. Children check in with their friendly listener when they get home. This provides a safety check and gives an opportunity to share feelings and ideas. Sometimes the children want reassurance about scary sounds, homework help, advice about personal problems or emergency situations. To ensure that the older listeners do not overstep their bounds, a training program is provided for the volunteers about their role and how to make referrals if necessary. Usually the listeners are available from 3–9 pm and their phone number is given to only two children. But this may result in as many as ten conversations a day.

4. *Extended school day programs.* In Brooklyn Park, Maryland, a poll determined that two-thirds of the elementary students were alone when they returned home from school. Now the parents who want their youngster to participate in an after-school program, from 2:45 to 5 pm, pay 5 dollars a week. The supervised experience includes a choice of sports, crafts, trips to the library, or tutoring. The program involves two staff members and a pool of fifty teenage volunteers. According to the kids, "It's more fun because we can do stuff together. Otherwise, when I'm home alone, mom doesn't allow any visitors." "I don't just sit and watch TV like before or overeat because of boredom" (Nolan, 1984).

ADOLESCENCE

Self-Discipline and Social Responsibility

Annual polls of the public attitude toward schools repeatedly identify adequate discipline as the foremost problem in education. Teacher surveys reveal the same verdict. Not surprisingly, there is diversity of opinion among teachers and parents regarding the meaning and exercise of discipline. We define discipline to be the maintenance of order in a school by both teachers and students so that conditions are conducive to the achievement of stated educational objectives. The ultimate purpose of school discipline should be to develop socially oriented self-control in students.

Recent years have seen a marked decline in the tendency to use corporal punishment at school or to physically restrain or remove students who violate propriety. Indeed propriety has become difficult to define. Parents, who were once supportive of stern discipline now condemn and even sue teachers who use physical force on disruptive, aggressive, destructive students. In most school districts unacceptable student behavior that is not grossly disruptive or detrimental to the class usually results in informal discipline as determined by the teacher. A few of the options available to the teacher are: teacher-student conference before or after school; referral to an on-campus ombudsman; peer review; detention; cooling-off room; sound-off board; enlisting services of community members such as minister, probation officer, doctor, or social worker; conference telephone calls with parents; case conference; transfer of student to another class or program; individual or group counseling (Hawley, Rosenhaltz, Goodstein, & Hasselbring, 1984).

Formal discipline results from, but is not limited to, instances of assault, theft or extortion, possession of weapons, use or selling of drugs, gambling, incorrigibly bad conduct, and obscenity. All cases of formal discipline are handled by administrators or the board of education utilizing due process. Unlike elementary school where individual teachers make up their own rules and decide about punishment as individual infractions occur, high school students are subject to a common set of rules and expectations. Moreover, these are not determined exclusively by adults. Typically a task force consisting of students, parents, teachers, and administrators develop the discipline policy which is then distributed as a rules booklet to every student and discussed in the homeroom as well as school assembly. No one can say, "I didn't realize this would be the punishment if I got caught fighting," or "I thought it would be possible to negotiate the punishment." All students are considered accountable for their behavior, obliged to exercise self-control, and expected to become socially responsible.

Self-Evaluation and Self-Control

Student awareness that misbehavior will bring external punishment is not the only way to ensure social responsibility. It is also worthwhile to encourage self-evaluation. While everyone should consistently experience a favorable self-impression, there are times when it is a person's best interest to feel ashamed of his behavior regardless of whether it leads to disapproval by others. Although this experience is a necessary part of growing up, students are seldom called upon at school to engage in self-evaluation or given guidance in developing the ability to do so. It is true that an increasing number of faculties have recognized the need to set aside a commonly understood place where students who are upset, for whatever reason, can go to let their anger simmer down or regain their composure. By allowing this practice of temporary retreat, the faculty confirms it as a necessary first step in dealing with negative feelings toward others.

It is assumed that students who choose to go or are sent to the cooling-off room, time-out room, escape room, or whatever the retreat location is called, will confront themselves in some rational way and make the necessary changes in their behavior. A better approach is to make known a recommended process for self-evaluation, discuss it during student orientation, and periodically assess its benefit by student polling. Consider the following guide for student use:

Time-Out Guide. Whenever we get into trouble, it's a good idea to ask ourselves some questions about what went wrong and how to avoid further problems. This kind of evaluation is an important part of growing up. When you take time out for self-examination, think about the following questions.

> *What did you hope to accomplish?*
> You are the only person who can answer this question. Try to **ANALYZE YOUR PURPOSE**.
>
> *Why do you suppose your way didn't work?*
> If something you're doing doesn't work, maybe you should look for a better way. Try to **ANALYZE YOUR SUCCESS**.
>
> *What are some other ways you could handle this situation?*
> This task may require a helper. Adults can't do this well by themselves. If we wish there is a peer counselor available or a retired adult volunteer. Try to **GENERATE POSSIBILITIES**.
>
> *What might be the consequences for each of your choices?*
> You can make better decisions by exploring in advance where each of your choices might lead. Try to **PREDICT OUTCOMES**.
>
> *What do you think are your best choices?*
> Try to **DECIDE A COURSE OF ACTION**.

This procedure can be modeled by the faculty, used by students in school, at home and throughout life. It supports the development of self-examination and social responsibility.

The Just Community

Like the advocates of values clarification, Kohlberg (1983) feels that moral development is hindered more than helped when adults try to impose values. A better method involves developing a natural dialogue among peers. In order to create a peer setting within the school, it is first necessary to radically transform the role of students and teachers. Only when all parties have the same status is it possible to eliminate coercion, obedience, and the expression of inauthentic behavior. By working together as equals it is possible to establish a "Just Community."

Kohlberg founded the Cluster School in Cambridge, Massachusetts in 1974. This alternative high school consisted of 30 students, 6 teachers, and a host of consultants. The participants were committed to the notion that adolescents can learn to make society a "Just Community" when they are allowed to influence their own school. Initially it was supposed that through group discussion, role playing, and due process everyone could easily make the transition to a democratic process. On the contrary, some students tended to depend on teachers to make decisions. Sometimes mob rule prevailed in the discussions. A few teenagers wanted power transferred exclusively to students. There were instances of theft, drug abuse, cheating, and absence from class. For each of these transgressions it was necessary to conduct town meetings. The process of democratic confrontation meant that consensus was often heard to reach. Gradually, however, more reasoning and less self-interest began to prevail. Although certain students continued to be easily influenced by group pressure, more of them became careful listeners and readily asserted their point of view.

In 1979 the project was completed and the Cluster School closed. Based on this experience Kohlberg altered some of his thinking about moral development. According to the amended theory, there are several stages of moral judgment. In the first stage people obey rules just to avoid punishment. This is the moral reasoning expected of young children. Later, as persons grow, morals are progressively based more on reason and ethical consideration. Kohlberg emphasized stages because his research indicated a sequential movement that did not skip stages. Individuals move at various rates and their moral development may be arrested at any stage. Although each step forward requires more mature thinking, Kohlberg does not claim a direct correspondence between cognitive development and moral behavior. Snarey's (1985) review of 45 studies offers strong support for Kohlberg's assumptions.

Since 1980 many school districts throughout the country have implemented some variation of the "Just Community" concept. These efforts show a respect for the opinion of students and recognize their need to share in institutional decision making. Perhaps the most ambitious program involves Brookline High School in Massachusetts. With the aid of a federal grant and Kohlberg's assistance, the 2,000 students, teachers, administrators, secretaries, and custodial staff are trying to improve themselves and their school by combining participatory democracy and moral education.

Early Adulthood

Developmental Challenges

The process of socialization expands with what might be called explosive force in early adulthood. Men and women have to move by abrupt steps (graduation, taking a full-time job, getting married) from a situation of dependency on parents and teachers to one of concern for others. Failure to make the changes on schedule results in weakened self-concept and alienation. Specifically, the young adult is faced with such socialization problems as:

> Adjusting to one's spouse as a peer—as an equal. This is in contrast to the one-way process of being loved and cared for by concerned adults that prevailed while one was growing up. Marriage partners are engaged in a reciprocal process.
>
> Teachers were concerned about helping students do their best. Employers are primarily concerned about getting the job done. Only to a limited extent are employers patient with the learning and improvement process.
>
> Children are imperious in their demands. Their lives revolve about physiological and safety needs. Maturing adults must provide for these needs without ostensible gratitude from the children.
>
> There are community jobs to be done. Some people

may ignore them and others may delay; but the need for a wide variety of volunteer services persists and grows larger.

The cross-age unity of socialization may be illustrated in the rapidly changing role of parents. Initially they must deny their own preferences for sleeping, recreation, work, and time with friends to satisfy the immediate and insistent demands of infants. Next they must focus on teaching young children those attitudes, values, and behaviors which society demands. Then they must relax the authoritative role in favor of becoming a confidant or advisor. Still later, and this task carries over into midlife, they must approve or perhaps even urge sons and daughters to move away from home. A similar evolving process occurs in relation to the spouse, employer, community, and church. Life requires that successful adults continue a process of self-examination and a process of self-improvement in each of their relationships.

Self-Concept and Identity

Ardrey (1970), a well-known social biologist, postulates that everyone has three innate needs, all of which demand satisfaction. As shown in the list, the highest need is for identity, the opposite of anonymity. Below identity is stimulation, the opposite of boredom. Security is the lowest need; if it is unfulfilled, the result is anxiety. It may be that in our society as large a proportion of the people achieve security and as small a proportion attain identity as ever before in history. This massive failure to ensure the satisfaction of our highest need can be explained in part by changes in the workplace, the nature of jobs in large impersonal organizations, unfair competition in schools and by life in overcrowded neighborhoods. Whatever the reasons, identity has eluded millions of persons for whom there is no rank, no territory, and no sense of significance.

Hierarchy of Needs	Consequences if unsatisfied
Identity	Anonymity
Stimulation	Boredom
Security	Anxiety

Achievement of security and release from anxiety presents us with boredom, the psychological process least appreciated by social planners. We are only beginning to see something of the chaos that a bored society can produce. It is more than coincidental that shock and sensation have arrived at the same time in most technological countries. The frustration of boredom triggers a high growth rate for pornography, fantasy drug trips, kinky sex, casual adultery, petty theft of unneeded goods, and violence for kicks, as well as street vocabulary and gratuitous horror in films and in the works of literary figures. These phenomena suggest that many people have remained fixed at the level of boredom/stimulation. Failing to achieve identity, they exploit the possibilities of sensation as their only escape from boredom. In this sense massive boredom accompanies being a technological people with little chance to develop and preserve a sense of identity. It could not exist if, along with healthy stimulation, we met the need for influence and respect which in animals is met by territory or social rank (Ardrey, 1970).

In some countries, people must settle for anonymity as the price of achieving security. In the United States, however, we are committed to enabling individuals to achieve identity (Erikson, 1980). We need not give up this goal but there may be good reasons for reconsidering how we go about achieving it. During the past few years the prevailing view has been that, to enable more people to feel important, we must create an increasing number of new job status titles and awards. Although this approach has in some cases met the need for recognition, in other cases it has made the quest of recognition insatiable (Wallach & Wallach, 1985). It seems obvious that job holding no longer brings the sense of satisfaction and identity that it once did. When people say "Thank God it's Friday," what do they mean? Do they mean that they achieve so much pleasure and extract so much significance and self-fulfillment from their jobs that the sheer ecstasy cannot be sustained for more than five days at a time? Or do they mean that their jobs lack importance for them, fails to satisfy their need for potency and forces them to look to the weekend for fulfillment? Do they mean that it would be nice not to have to hold their jobs but that the need for income and social status demand it? The inner turmoil of millions of workers does not signify that they are lazy or that constructive achievement doesn't interest them. It means that alienation from the job is becoming more common.

There is a more promising way to attain widespread identity in a leisure-oriented society. We can make identity possible for a greater number of people by removing the barriers to their involvement in the expressive culture (those activities in which people participate for enjoyment of the activity rather than for some purpose beyond the activity). The government began to support this view a few years ago when it served notice that females in college and high school should no longer be denied access to team sports. These and other legislative efforts to reduce the scope of sex and economic discrimination are commendable. But even laws that guarantee equal opportunity will not greatly increase participation in the expressive culture so long as people are led to exclude themselves. This is what happens when they are taught to regard the outstanding performers of an activity as the only ones who deserve to participate.

The use of a restrictive standard for determining who is worthy to participate begins early in our culture. It can be observed in little league sports. Similarly, within the classroom, teachers feel obligated to find out what a student is "good at" and encourage tasks calling for that particular strength. In effect, the message is that persons ought to limit themselves to those leisure pursuits in which they excel. Unless someone is identified as having extraordinary talent in music, art, or some sport they should withdraw because the quality of their performance will not support a favorable self-impression. Under these conditions some people feel compelled to give up certain activities that could provide them with lifelong satisfaction.

MIDDLE AGE

Emergent Perspectives: The Need for Choice

Social obligations continue in middle age. However, the demands of younger members of the family, spouse, and work supervisors may be less strident and less pervasive. Hence there is a temptation to ease off, to let matters slide. This may be compared to one's place in highway traffic. One pulls off to the side to take a breather. In England places for this are provided in what are called "laybys." In our country there are "Passing Zone Ahead" signs, safety zones, and rest stops. Drivers in England actually stop to make a cup of tea. In the United States, on holidays free coffee is offered by the Lions Club and other safety-minded organizations. But the mainstream traffic still speeds on. Nonetheless these slowdowns are regarded positively by traffic experts.

One can choose to stop, or slow down, reflect, and change pace in midlife. Because persons have gone along with the traffic they may have become lopsided in goal choice or pace of living——focussing too much on occupation, family, or community to the neglect of keeping all facets of socialization in good working order, in balance.

Psychologically and sociologically, regardless of personal views such as "You're only as old as you think you are," persons in midlife should also be developing new outlooks. Precisely what these new perspectives should be are not a matter of consensus. Roberts and Backen (1981) report that there is an improved climate for discovering some of the particulars in what is called "agenda setting." This refers to the ability of the mass media to influence the level of the public's awareness of issues as opposed to providing specific knowledge about these issues. Persons preparing materials for media documentaries may not have solved any problems but they have highlighted certain aspects of lifestyle and issues of role change to which middle age adults should give attention. These are not necessarily novel concerns but magazine articles and television special productions have increased awareness of them:

> Spouses need to be concerned about relating to their mates as persons rather than as sex objects or parents.
>
> Attachment, being a salient feature throughout life—not just in infancy—must be dealt with in marriage, friendships, and vocational pursuits.
>
> As persons encounter crises such as loss of significant others, physical impairment, monotony, and loneliness, there is an alarming increase in the search for chemical comfort; alcohol, Valium, Lithium, and so forth.

Caring for Aging Parents

Adults need to shift to, or include with their children, concern for aging parents. Elsewhere we have discussed the relationship of middle-aged persons to their spouse and teenagers. Here we consider yet another role, one for which few people are adequately prepared.

Eighty percent of elderly people have adult children. Most of the parents live in their own homes, usually within an hour's drive of a daughter or son and see them at least once a week. Usually the generations support each other as needed but with advancing age elderly parents eventually require more care than they provide. Although this circumstance is predictable, it is rare for adult children to know how to deal with an enlarging caretaker role.

The communication of needs is vital for a good relationship between aging parents and adult children. It is important to reach a mutual understanding of what parents need and how the adult child can make a contribution. Both parties should make known their view of the parent's needs. For example, mother and dad may initially feel they need weekly help with housecleaning and yard-care. Perhaps later the list will include transportation to the market or church. They may also want some designated visiting time to spend with grandchildren. But, unless these needs are expressed, the necessary assistance may not be forthcoming. It is unreasonable for the parent to assume that a son or daughter ought to anticipate every need without being told.

Conversely, aging parents may have needs they do not recognize but adult children can identify. For example, it is wise to focus most family conversations on current affairs so that everyone feels obligated to keep up with the news and express their own interpretation of what is going on. This emphasis on the present time keeps parents from excessive reminiscence, from talking only about the past. In this connection, some men and women recognize that their parents are inadvertently becoming isolated because they have insufficient socialization outside the family. When this realization occurs it deserves discussion with the parent. Children should also be willing to make known the parental needs they see for continuous learning, volunteer service, and growth as a grandparent.

It is a mistake to suppose the only person with needs is the aging parent. Adult children must determine what they can reasonably expect of themselves so they can remain an effective source of help. Caring for an aging parent can be very stressful. Generally the primary caretaker is a daughter who is also a working mother. In combination her responsibilities to spouse, teenager sons and daughters, employer, and aging parents can be overwhelming. It is little wonder that some women feel physically worn out and emotionally exhausted; others say they feel tied down by a daily care schedule and resent having to give up their preferred social and recreational lifestyle; still others report being discouraged because their parent is never satisfied no matter how much effort is expended. In a study of stress among elders and caregivers in Michigan, it was found that the adult caregiver was three times more likely than the elder to report symptons of depression and four times more likely to report anger (Gelman, 1985). When this kind of stress is sustained it can undermine the adult child's marriage, lead to personal breakdown, or result in elder abuse. The number of elder abuse cases per year is estimated to be as high as 2.5 million (Brubaker, 1985).

The need for periodic relief from looking after parents implicates siblings and community agencies. It is appropriate that the adult children who do not regularly provide care to their parent arrange to give siblings some relief from time to time. Some sons and daughters who live in other cities may elect to fly the parent to them and provide care for a designated period. In other cases they can come to stay with the parent while sister or brother take a vacation. Besides sharing the load with siblings, it is wise to become familiar with community services that might lighten the family load. For every person in a nursing home or long-term care facility, two remain in their homes receiving community help like Meals-On-Wheels, housekeeping aid, daily reassurance calls, frequent church visitors, adult day care, dial-a-ride transportation assistance, and socialization activities at the senior citizens center.

In the future, society will have to devote more resources to the care of aging parents. The greater emphasis is justified because of increased longevity, larger numbers of older people, more mobility which separates families, increased employment of women (the traditional caretakers), a growing rate of divorce and higher incidence of single-parent families whose ability to provide care for the elderly is diminished. It will also be necessary to devise and offer commonly available educational programs for adult children and aging parents alike so that they can better understand one another and more often achieve the family relationships they desire.

Later Adulthood

Peers and Retirement Communities

Being a burden inconveniences adult children and lowers self-esteem of the elderly. Therefore, it can be predicted that, the stronger the desire for family respect, the greater the older person's motivation to remain independent. Under these conditions, it is not surprising that approximately 75 percent of the elderly choose to live away from relatives. An increasing number of retirees are choosing to live among peers. As a result there has been a tremendous growth in retirement communities, especially those located in the climate-favored states. While young adults believe that life in an age-segregated community must be uniform and dull, surveys show that most older adults would prefer to live in a community inhabited largely by people their own age (Dentzer, 1985). To account for these contrasting perceptions, we must bear in mind the loss in roles, norms, and reference groups that accompany aging.

During childhood, boys and girls are encouraged to "act your age." This advice implies that, at least for the initial stages of life, there are predictable sets of expectations. But once beyond adolescence, culture and tradition demands that people substitute their occupation and the relationships required there as a normative base of expectations for behavior. Still later, when age dictates retirement, many individuals find themselves for the first time lacking a reference group for their behavior. To be without norms is to be alone in the worst sense, because one does not know the proper criteria by which to evaluate oneself. This is especially critical during later life, when what little one has learned about aging supports fear. Thus, to be old and lack norms is to wonder: "Do other people my age have the same problems with their bodies? Am I different or about where I should be for someone my age? Am I healthy or sick?" For many of the elderly, the need to use a physician as a substitute source of norms is shown by their high frequency of office visits where there is no evidence of organic disease.

For certain people, the consequence of role and norm loss can be fatal. This danger was first recognized a century ago by the French sociologist Emile Durkheim (1952), who set out to determine what causes people to commit suicide. When his analysis was complete, Durkheim suggested that the major cause of suicide is intense loneliness, coupled with a lack of clear expectations resulting from a marginal social position. We know from more recent research that suicide is often the consequence of a loss of roles, a loss of norms, and a loss of reference groups. It is more than coincidental that suicide rates beyond adolescence show a larger increase after age sixty, especially among men, than for any other age group. Miller (1979) studied all the cases of older white men who took their lives in Arizona over a six-year period. The results pointed to role and identity losses at retirement as having a significant influence. It was also learned that physicians and relatives must become better able to detect and respond to signs of depression. For example, Miller observed that three of every four suicide victims in metropolitan Phoenix had visited a physician within a month before death; 60 percent of the victims presented verbal or behavioral clues to their family. But these indicators either went unrecognized or were not taken seriously.

If normlessness can lead to self-destructive behavior, how can we compensate for the inevitable loss in roles that people will continue to experience upon retirement? Gerontologists have contended for some time that age-segregated housing is one answer (Rosow, 1967). They argue that older people in this country are disadvantaged by their lack of roles and normative expectations. Because they have few recognized functions and fewer norms to guide their conduct, one way for increasing norms is deliberately to create an age reference group by means of residential concentration. In bringing the elderly together, the retirement community can be valuable for socialization into old age and for the development of age-appropriate expectations.

The Grandparent Role

One role that has attracted considerable attention in the past decade is grandparenting. This interest is revealed by the elderly, by both nuclear and extended families, and by hospitals, schools, and homes for the aged. In part the concern stems from traditional concepts of grandparent roles being replaced by professional advisers and surrogate parents. The diminished importance of grandparents fostered by social change and greatly increased mobility of extended families has placed older people in an ambiguous role. They are assured that they are needed but in practice virtually ignored. Families do a disservice to grandparents by alleging their importance and then failing to use their talents.

Parents have access to self-help books and classes to improve childrearing expectations. Similar opportunities should be available for grandparents. Instead they are left alone to wonder "What are my rights and responsibilities as a grandparent? In what ways can this role be more influential and satisfying? How well am I doing as a grandparent?" These kinds of questions will persist until norms of constructive behavior are established for grandparents to use in goal setting and self-evaluation.

At Arizona State University a curriculum for grandparents has been designed which can help them understand their rights, roles, and responsibilities (Strom & Strom, 1985). Grandparents who participate represent various ages, ethnic backgrounds, and income levels. They come together weekly to learn about sharing feelings and ideas, asking children questions, improving storytelling and acquiring skills in self-evaluation.

Grandparents want to think well of themselves. But the respect they desire is difficult to attain because it depends on fulfilling a relatively undefined role. They aspire to success but lack a common set of reasonable criteria for self-evaluation. Grandparents need norms of constructive behavior so their roles can become more influential and satisfying. Our method for focusing self-evaluation begins with grandparent groups brainstorming a list of the rights and obligations they feel are appropriate for themselves. Later the merit of their choices is examined in group discussions.

The need for social adjustment in later life implicates all generations. Regardless of where they live, grandparents—real and surrogate—may profit from participating in educational programs which help them understand their rights and obligations, acquire skills for getting to know grandchildren, realize how childrearing practices have changed, and determine ways to improve their families and intergenerational influence.

Chapter 9

FOSTERING EFFECTIVE COMMUNICATION

For several decades scholars have questioned the traditional concept of intelligence that confines mental activity largely to language and number. Many authorities want to have a broader base for intellectual assessment than the current emphasis on language and problem solving. The plea is to include a greater range of human behavior—creativity, social interaction, aesthetics, and motivation.

Because language does have a heavy impact on thinking and social interaction, it will remain a major facet in the study of human development. Speech is the audible manifestation of intelligence; it is a distinguishable part of social personality. Our speech more than our face or body reveals our personality to others. Because speech is a unique part of being human, and because thought is mainly a matter of talking to oneself, improving the ability to communicate is an important aspect of human development. Certainly health and emotions cannot be ignored. But because so much of behavior is linguistic, or is conditioned by it, communication demands a star role.

In this chapter we (1) identify the aspects of communication that are most prominent at various age-stages, and (2) recognize the processes which promote optimum communication, reciprocal dependence, and cooperation throughout the lifespan.

The Meaning and Role of Communication

The Concept of Communication

Communication is focal in marital happiness, neighboring, family accord, instruction of the young, and in most work settings. Most futurists agree that communication will be even more important in the information era which we are now entering. According to Hofmeister (1984):

> We represent the last generation of the industrial age. The pupils presently coming to school are the first generation of the information age. When our present eighth graders take their place in society, 75 percent of them will be involved in information related industries. We are participating in a massive change in the very structure of society. For those of us whose lifespan will include the transition between these two ages—the industrial and the information age—this is indeed a time of wonder, challenge, and confusion. Like the adolescent caught between childhood and adulthood, we are experiencing that strange mixture of excitement and confusion as some of our traditional reference points dissolve and we try to determine which of the new directions has substance and which are shallow, seductive facades.

It would seem that communication, which is almost as natural, common and lifelong as walking, requires no definition. However, analysis shows that communication is a complex and intricate process that is attracting, if not demanding, a considerable amount of current research.

Cultures, even in ancient times, gave some attention to communication. For instance, the Chinese Mandarins maintained their ruling class privileges by speaking in one language to their peers and using another when they talked to lower class persons. In theory the ruling class was open to any who were able to assume responsibility. However, to gain admission to the ruling class civil service examinations had to be passed. These tests were given in the Mandarin language. Thus, the entry of outsiders to the ruling class was blocked. Witch doctors use one language when talking to their gods but another when they speak to tribal members. The witch doctors are not much different from auto mechanics who use an uncommon language when describing the malfunctioning of a car. In short, speech may be used to confound communication. (In fact, over time the professional fields have each developed their own jargon, their own language, that often is little understood by outsiders.)

Considering tone, emphasis, inflection, and vocabulary, fathers use one language when addressing members of the board and another when talking with their children. Similarly, politicians use one language when designing treaties or party platforms and another when they explain their activities to constituents. But it is only in this century, with the work of semanticists, that scientific study of metacommunication (the nature of the communication process) has become serious. We are becoming more aware that there is a difference between talking and communicating. In fact, talk is sometimes used to obscure communication. Linguists call such speeches semantic blanks because nothing has been communicated.

The Vital Importance of Communication

There are scholars who assert that we are emerging into another kind of socioeconomic world—one in which communication is focal. This perception forces us into a novel appreciation of the meaning and importance of communication. Naisbitt (1982) refers to this new socioeconomic milieu as the Information Society. We have moved from ancient nomadic life through the agricultural age, through the industrial production culture to the society in which information-communication is the means by which many persons make their livings. The emergent society is not just service-oriented (as some scholars had predicted) but oriented toward information and its transmission. This is difficult to comprehend even as we think of television, computers, and satellites. But when we consider those who work as programmers, teachers, clerks, secretaries, lawyers, and politicians we can begin to appreciate the size of the communication arena.

The trend of our culture toward bigness and anonymity and toward crowds and masses is viewed by some authorities as a threat to social survival. Rogers (1983) suggests that our greatest survival problem is how much human beings can accept, absorb, and assimilate the rapid rate of change which characterizes our culture. The population explosion brings with it not only crowding but an increasing lack of intimacy. Rogers regards interpersonal process groups, basic encounter groups—the attempt to communicate intimately—as the greatest social invention of the century. These group processes are steps toward intimacy—steps toward identity. If they are used wisely the result can be less loneliness, better ways of resolving conflicts, and more satisfying personal relationships.

Bruner (1983) postulates that the processes of development, especially in infancy and early childhood, depend on sustained care, love, affection, and attention. The manifestation of these human qualities is in some degree linguistic and communicative. Bruner is a proponent of the activity approach in children's education; children need to be actively engaged (as contrasted to passive listening) in their own learning. A major approach to cognitive growth is to engage children in dialogue. Bruner contends that "The courtesy of conversation may be the major ingredient in the courtesy of teaching." In recent years there has been a growing emphasis on student participation, students' talking, and teacher-pupil dialogue. Of course, dialogue should be more than a school enterprise. Parents' talking with children about what has been read is recognized as a facilitator of reading skill. Similarly, talking about numbers, arithmetic in daily life, and matching objects (people and chairs, guests and table settings) helps children bridge the gap between abstract counting and numerical relationships in operation.

Maslow (1970) emphasized the role of communication in his description of basic human needs. The need for safety (reliance on rules and dependable persons), the need to love and be loved, the need for esteem and respect all are supplied in some part by communication. In fulfilling the need for self-actualization, communication is a facilitating factor. Basic encounter groups can help persons define, clarify, and project their perception of what self-actualization means and the forms it may take.

Yankelovich (1982) says the theory of Maslow, and others of the humanistic orientation (e.g. Carl Rogers, Rollo May, Erich Fromm, Gail Sheehy), has led to a "me first" ethic. He postulates a conflict, between an ethic of self-fulfillment (the present generation), self-denial (the ethic of our immediate forerunners), and the coming generations' new ethic of commitment. He does not see the ethic of self-denial as being all good. For example, some persons have suffered guilt as the result of enjoying life, prosperity, or being proud of their offspring. Somewhere between the hazards of self-denial and self-fulfillment lies an emergent ethic—the ethic of commitment. And this brings us back to the burden of this chapter. The fulfillment of Maslow's needs may be facilitated by communication and dialogue. The new, emergent, ethic and value of commitment, as postulated by Yankelovich (1982), may be clarified, defined, and put to the test by communication and dialogue. Conventional, outmoded, or emergent ethics must be verbally communicated, weighed, and shared before they are put into action (Wallach & Wallach, 1985).

Levels of Discourse

Most of the time, throughout this chapter, communication will refer to concerned, emotionally involved, discourse designed to promote interpersonal understanding and to decrease interpersonal distance. In order to clarify this idea reference is made to levels of discourse ranging from non-communication to that involving emotionally satisfying communication.

1. The lowest level (1) is neither harmful nor informative. It refers to the dialogue we use to initiate conversation or superficially acknowledge the presence of others: "How do you do?" is said without a smile or eye contact. "What can I do for you?" says the accounts-receivable clerk while continuing to type or dial. She speaks in the same tone as into the telephone. Even the cheerful, "Have a good day," spoken as she enters the amount paid, does little to erase the nonperson status she had awarded. The impersonality of level-1 discourse is communication only in the sense that words are used. Children's communication is inhibited because they are so egocentric as to believe that what they say is immediately understood. Adults (e.g., the clerk) as well as children, must learn that how they act, as well as the words they say, are factors in communication. Impersonal behavior increases interpersonal distance whereas the role of communication is to decrease interpersonal distance.

2. Factual, informative discourse (level 2) occurs when a person asks for and is given street directions. This is the level of discourse that characterizes, some, perhaps most, of instruction in the classroom or verbal transactions in the office. There is minimal or no emotional load. Some college physics or mathematics teachers may not care whether students like or dislike them or the course content. Someone teaching classes in group dynamics or principles of counseling probably has somewhat more concern with interpersonal distance and the therapeutic nature of communication than does a chemistry instructor. Elementary teachers, it is hoped, have much concern for interpersonal factors. Most educators are concerned about the motivation to learn as well as the content of learning; and they believe that part of the motivation stems from teacher-pupil rapport. These teachers are moving toward level 3. There is ample

evidence that at all levels of schooling optimum learning does involve motivation, teacher-pupil rapport, and pleasant physical atmosphere—in short, learning climate.

Level 2 discourse is the predominant type of communication. It is factual, logical, reasonable, informative, and utilitarian. Transactions with others can be conducted dispassionately, reasonably, and intellectually. Quite possibly the meaning of some information might be misinterpreted if effort is devoted to making it appear personal. Occasionally, especially in the face of disagreement between information sender and information receiver, it would be well to think of the next two levels.

3. Level 3 discourse concerns interpersonal communication and understanding. Its function is to decrease distance between persons. "I'd like to know where you are coming from," says the construction boss who wants to know why the tile layer is proceeding as he is. In short, "What is his motivation?" The question, sincerely asked and honestly answered, can be ego-satisfying and identity-building. When persons are communicating on level 3 they are expressing feelings about surroundings, they are discussing meaning and the significance attached to objects and persons other than themselves. Level 3 discourse is concerned with reality testing and the validation of perceptions. Participants are comparing and checking feelings, actions, and attitudes with each other. They are thinking through, adjusting to others, to present conditions, and to pertinent antecedent occurrences. It is an integrative type of communication which promotes psychological health and effective interpersonal relationships.

4. Level 4 discourse moves a little closer to the inner being of self than does level 3. Both are personal and emotional but level 4 exposes more of one's usually secret self. It takes unusual trust either in self or another to discuss one's private longings, fears, and self-perceptions. Psychological distance is minimal. Four is the level of discourse needed to deal with the alienation and anomie which characterizes so many people in our population. The potential school dropout—or life dropout—delinquents, criminals, unemployables, drifters, are portrayed as having little ability to conduct level 4 discourse.

Erikson (1980) refers to the alienated person as being rootless and thus having no sense of identity or attachment. Elderly men and women may feel alienation as the result of being ignored by their children or when removed from their homes and the stream of life. They need level 4 discourse to deal effectively with such feelings. The impersonality of management-employee relations can also contribute to work alienation. These are the concerns of level 4 discourse. The ability to be caring can be learned. The starting point is to talk about caring, alienation, anomie, and what it means to be open, frank, and honest. Level 4 discourse is not always a comfortable area. So, there is likely to be back and forth movement between confiding in others and a reluctance to do so (level 3). However, it is only when persons come to see themselves in relation to the environment and others, to see alternate and better ways of living that growth toward personal development can be sustained.

5. Level 5 discourse is achieved as a part of effective personality maturing. It is the communication that requires no words (or very few of them) to express the feeling of oneness with the world, another, or others. It is a level of discourses analagous to what Maslow (1970) calls peak experience. Words are quite inadequate to describe the surge of feelings some persons get from walking through a grove of giant redwoods, watching a sunset, or seeing one's beloved spouse sleep or smile. One gets this feeling of all's-well-with-the-world when one sees the bubbly grin of an infant. The feeling is shared with another by a glance or tightening of handclasp. Level 5 discourse is not the result of direct pursuit or demand. One can create the favorable psychological and physical environment. Persons can talk about it and seek to foster the verbalization and interpersonal relations that may preface its evolvement. Those who are confident about themselves and capable of trusting some others experience level 5 discourse—not frequently—but more often than those who have less self-and-other trust.

INFANCY

Amazing Skills of Infants

It is difficult to believe that our educated ancestors thought of infants as helpless, undeveloped, and inadequate as is reported in early books on child development. In contrast, recent research credits infants with an amazing repertoire of capacities and abilities. It seems they are very competent socializers and communicators (but not linguists). Mothers do not need to have child psychologists tell them when their infants are hungry, tired, happy, or lonesome. One mother reported that her baby always enjoyed her bath. But one day baby Emily screamed when she was put in the tub. Mother thought it was gas. She took her out, burped her, put her back in the tub. Emily screamed again. But mother continued. Finally Emily settled down. In the meantime mother noted that the warm water had been running all the time. The water had heated up at least five degrees. Emily wanted her bath but at the proper temperature. Mother said, "In fairness to Emily, the process of knowing her has been mutual communication—and the only secret is paying attention."

As Trotter (1983) put it "Children do not speak adult and adults do not speak children." Adults need to make the adjustment to child language because it will take two to six years for the child to speak adult. However, there are now findings to the effect that infants as early as six weeks can express feelings quite accurately. By the first birthday they have a range of ten to fifteen reactions that can be read accurately by trained observers. The key to understanding is observation (Sparling & Lewis, 1984; White, 1985).

Language Development in Action

Theories of language will hopefully yield practical clues to optimum linguistic development in young children. For parents, teachers, and other caretakers enough is known today for them to proceed with confidence. (1) Children need a respondent—a sounding board—to their vocalizations. (2) Children need noncritical feedback. (3) Children need an adult model.

(1) *Children need a respondent.* As we accept the proposition that humans are uniquely programmed for language acquisition, we should not ignore cognitive psychologists' plea for a learning environment (Gagne, 1985). Because some of the baby's vocalizations are spontaneous does not mean they can be left alone linguistically. Brain scientists and developmental psychologists should work together. Experimental data for the past half century have shown that despite spontaneity, if babies are left alone they soon stop vocalizing, developing, and even stop crying. The baby's respondent inevitably teaches which of those initial spontaneous sounds are needed in the speech of the child's culture.

The need for various sciences to collaborate in research on language is underscored by the title of an article, "Psychoneurobiochemeducation" (Krech, 1969). Krech's major emphasis was on the fact that the exercised brain had more glia cells, larger blood vessels, heavier cortexes, and was metabolically more active than the brain of children left alone. But, most of all, he emphasized that the brains of young humans need to be exercised by the "species specific" stimulus of language—bombard the child with words. This, most likely will be done by mothers, but fathers, siblings, grandparents, and other relatives and family friends can contribute.

Katz (1977) seems to be in fundamental agreement with Krech when she cites seven basic needs of young children. These are (1) sense of safety; (2) self-esteem; (3) experience of life as being satisfying and interesting; (4) help in clarifying their experiences; (5) adults who exemplify authority; (6) association with adults and older children who exemplify a whole lifestyle; and, (7) experiences with adults who take a stand on values. Reflection on these shows that communication is especially prominent in #4 and #7, quite significant in #2 and #3, and has some role in the remaining needs.

(2) *Children need to have noncritical feedback.* It has been noted that there is an almost inevitable tendency on the part of adults who talk with babies to instruct, expand, and improve (Chomsky, 1972). However, care must be taken that this not be done in a critical, "No, no," or "Try it again," fashion. Because of criticism some children cease to try, some will vocalize only to dolls, and some begin to stutter as they attempt to conform. In this connection, various cultures differ in the emphasis placed on language acquisition and correct usage. Some show much patience. It seems that adults in such cultures depend on time and maturation to aid the child's language acquisition. The Ute Indians (an American tribe) do not even have a word for stuttering, and there are no stutterers. In American culture, where matters of form and pronunciation are more important about one percent of children are stutterers. Linguists advise parents and teachers to be less hasty and insistent in making grammatical and pronunciation corrections. Identification of a child as a stutterer should be done only with caution and professional advice.

(3) *Children need an adult model.* The use of babytalk probably does more to retard, than it does to enhance, acquisition—except that the child does have a respondent. It is interesting and amazing to observe and listen to boys and girls as they acquire new vocabulary. Adults need have no fear that they will overwhelm the child with big words; unless implicit criticism, sarcasm, or disdain accompanies the response to the child's admission of not understanding. Typically, the child will ask for an explanation or will make guesses (usually with some degree of accuracy), or will focus on those parts of the conversation that are understood.

Although it is possible to communicate feelings and thoughts with a minimum of vocabulary, the more words at one's command, the richer and more exact the speech. Each person needs language facility to express ideas, to label thought, to urge the consideration of feelings, to describe emotions, and to compare experiences. Everyone has experienced trying to convey an idea when the appropriate words seemed fugitive. The problem is more acute for persons whose speech has developed in settings where a minority linguistic system or a restricted language is prevalent. People from these backgrounds often find themselves less able to understand and to make themselves understood than peers who come to school with facility in the language of the dominant culture. The greater the access to vocabulary, the less frequently all these frustrations occur.

Early Childhood

Family Televiewing

People of every age group enjoy television. They are motivated by personal interest. It would seem that a good place to begin to improve family communication is with television, the most common activity in which parents and children spend extended periods of time together. According to a Gallup (1985) Poll, three out of four parents regularly watch television with their children. When full advantage is taken of this opportunity for interaction, a better relationship can be expected.

The effects of television include both good news and bad news. Some programs for young children such as Sesame Street and Mister Rogers have been found to offer considerable benefit. It has also been determined that televised violence can lead boys and girls to sanction aggressive behavior. The explicit sexual activity as well as violence that children view is expected to increase as more families have access to cable programming. This prospect has brought about the formation of child advocacy groups whose intention is to favorably influence the content of television (Singer & Singer, 1983).

Given the potential for youngsters to see inappropriate events on television, it is not surprising that many parents have come to regard themselves as a censor or judge, someone whose age and experience qualifies them to decide what programs are likely to support or undermine the development of wholesome attitudes. In some cases, particularly within the families of 10 million latchkey children, parents have felt obliged to purchase a control device which permits them to lock out any unwanted programs that could otherwise be viewed in their absence. Certainly it is important for mothers and fathers to monitor what children observe. But besides being a censor, grown-ups should also assign themselves an interactive guidance-oriented role.

During televiewing parents and children look at the same pictures and hear the same words. But previous experience causes them to sometimes reach dissimilar conclusions. It is these differences in perception which enable family members to benefit from one another. An effective way to find out how a child interprets television is by raising questions. When parents take this initiative, it shows they care about what sons and daughters think and are interested in understanding them. To be more specific, the following questions can be used while watching almost any television program.

How would you handle the situation he's in now? The ability to identify alternatives is an important asset. People who can see many possibilities in a single situation are more able to negotiate, get along with others and think of options. These strengths serve us well in problem solving, conflict resolution, and preservation of mental health. By sharing viewpoints, parents and children reveal the scope and the limits of their individual perception.

What has happened in the story so far? The need for sequence recognition begins at an early age. One of the fundamental goals in learning to read is comprehension. Memorizing the alphabet and being able to identify words is not enough. Understanding the sequence of events is also important. In the books from which children learn to read, the words are necessarily short and the story line is uneventful. On the other hand, television programs usually have a beginning, middle, and end. Indeed most of them are better suited for the assessment of understanding sequence than are beginning children's books. It is therefore appropriate to ask ourselves: Can we teach an important reading skill by means of another medium than print?

How do you want the story to end? The expression of personal preference reveals our values. It is common for opinion polls to regularly report the desired future of adults, the way they want things to be. But the preferences of children are seldom assessed. This is unfortunate because the future we hope the young will enjoy depends in part on helping them now develop a sense of what is possible, an attitude of optimism, and a willingness to express their values.

Do you think he is making the right decision? In this case the goal is to evaluate judgment. Grownups want children to use good judgment and refrain from making serious mistakes. Sometimes harmful consequences take place before these lessons are learned. By considering the televised versions of real life dilemmas families can simulate problems and determine the worthwhileness of personal judgment without risk.

What do you suppose _____ means? Learning the definition of new words is a lifelong task. Many of the words that children hear on television can be the focus of this question, e.g., witness, victim, emergency, delay. An emphasis on vocabulary building calls for the more informed of the viewing parties to define words. It is also helpful to provide corrective feedback when statements reveal misunderstanding.

There is much to learned about how to creatively use television for supporting family communication. One way parents can make an important breakthrough is by expecting some new things of themselves. In particular, they should arrange to watch television with their children—this takes time. Next, they should by asking questions of their children during these mutual observations—this is a skill which takes practice. And just as parents want children to make known their impressions of what they have seen, mothers and fathers should also share their own experiences with them—this requires self-disclosure. Finally, parents should allow sons and daughters to choose some of the programs that the family watches together—this calls for acceptance of children's interests. Parents who subject themselves to expectations such as these communicate more easily with children and establish themselves as a lasting source of guidance (Strom & Strom, 1986).

MIDDLE AND LATER CHILDHOOD

School Life and Communication

Teachers who listen. Traditionally school is thought of as the designated place where children acquire the basics for learning to learn. Only recently have we begun to see that function in terms of a communication challenge. Not only is the transfer of knowledge a communicative process but the teacher-pupil rapport, which facilitates learning, is also a matter of communication.

Primary teachers set the stage for the trust that students need to have before they can confide their feelings. An atmosphere of safety is essential. Students must know that what they say will not be held against them. Secrets will remain secrets. Grammar is not always the most important item in verbal exchange. This atmosphere is as much an attitude as it is a matter of verbal assurance. The "heart" listener is genuine. It is not necessary to play roles or wear facades. Either words or an attitude of concern can convey the message that the student does not have to cover up or build a wall. In effect, the teacher is saying, "I will not look down on you. I accept you as is."

Many teachers need to improve their own listening skills. This viewpoint is supported by a nationwide study of 1,230 families representing those with children under age thirteen. One of the topics discussed with the participating elementary students was "ease of communication" with significant people in their lives. It was determined that, next to their friends, children find it easiest to communicate with their mothers. Next came sisters, brothers, and fathers. The people with whom the most children find it hard to communicate are school principals and teachers (Yankelovich, 1977).

Teacher-pupil-parent conferences. The essential nature of listening and communication in children's school life is captured in the practice of three-way conferences. Usually parent, teacher, and student come together only if there is trouble. Viewed from a different perspective, however, the three-way conference seems to be a good idea for families that are willing to share accountability. Here are some of the advantages:

> Children are encouraged to set some of their own goals.
> Children believe their input is important and valued by adults.
> Children collaborate in partnerships with their parents and teachers.
> Children's progress can be assessed from more than a single perspective.

Children's reliance on peers for self-assessment is reduced.

All parties hear the same information.

The main purpose of these teacher-pupil-parent conferences is to make each child an involved participant in his own education. This means cooperative goal setting, and bridging the gap between home and school life. In these conferences all parties get to know each other as they listen, observe and talk. One obstacle to such conferences is lack of time—both for preparation and conduct of the meeting. In addition, teachers need training in group processes so that no one ends up in the role of being a listener only.

Group Processes. The idea of Donaldson (1978) that adults tend to underrate children's reasoning power was illustrated for us during an observation of third graders while they were discussing a matter of discipline.

> Todd, one of the larger boys in the class, was hyperactive. The children were seated on the floor discussing the conduct of this boy—while he rolled back and kicked his feet in the air. Students remarked that he bumped others, claimed attention, and in general disrupted class procedures. He had been removed from the noontime movie because he was bothering other viewers. The eighth-grade boys, acting as monitors, had taken away his ticket and reported him to the principal. Some students were sympathetic. They felt the boy should have been warned. It was all right for him to be kicked out; but, "He paid for his ticket," so it should not be taken away. The class finally decided that they would have their representative go to the principal and request return of the ticket. Todd then could, on the promise of good behavior, see future movies. One little curly-haired blond boy said, "I don't know if we should do that. When Todd makes a disturbance it is just like he was taking away the tickets of the other kids."

That was the end of the discussion. But we were intrigued by the discussion and later returned to ask the teacher what happened. She said that everything was under control. Todd was a real problem, which the father recognized. The father had come to school and asked the teacher to try to put up with Todd. His mother had left the family but they were gradually settling down. The teacher told Todd that she could not have him in class when he misbehaved. It was decided by all three parties that when Todd could just not be still or act calmly, he would be excused for a while. Sitting in a chair just outside the room, he would be allowed to come back when he calmed down. Todd has learned to say to the teacher, "I'd better go outside for a while." It worked. We have in this one case group discussion, teacher-pupil-parent conference, the process of listening, and the wisdom of the young—very young.

Learning to Read

Reading is an intergenerational challenge of amazingly large proportions. It is estimated that there are 20 million English-speaking, native-born American adults who read so poorly that they have trouble getting and holding jobs or they suffer from marked loss of self-esteem (Wallat & Piazza, 1986). Those adults who read only at the third- to fifth-grade level are called functional illiterates. The handicap becomes increasingly great as automation, computers, and the information-communication society displace routine jobs. Help is available for these people, with increasing reliance on those who volunteer their services. Fortunately, because they have adult vocabularies, motivation, and experience, these illiterates often make remarkable progress with help.

Ideally, this potential handicap (and it is an especially good word in the communication area of development) should be overcome by parents, aunts, uncles, cousins, grandparents, and volunteers who read to preschoolers. At home or in classrooms for the culturally different such reading can help generate the interest in books and reading that will ease the task of teachers who must daily face 20 to 30 "individual differences."

Typically, teachers do a commendable and conscientious job of teaching children to read. Their task is made much easier when students have experienced the phenomenon called readiness. Basically this means that a given child through physical and mental maturity and experience is prepared to tackle successfully the intricacies of reading. Obviously the youngster must have the vision to discern the word in the book or on the chalkboard. One in every four school-age children is affected by eye problems. In many cases, it is not until they approach reading that some students are found to have astigmatism, are nearsighted, or farsighted. Some have the slight brain damage that reveals itself in letter reversals; e.g., they cannot consistently distinguish between *how* and *who*, or *arm* and *ram*.

There are many contributors to reading handicap, not all of them physical. One contributor worthy of mention, because it can be readily controlled, is the "tyranny of the average." This is the fallacy that students should at least keep pace with some statistical norm. If, by Christmas or the Spring break, first-graders have not mastered their initial attempts to read as readily as their peers they tend to become "cases." Terms like nonreaders, retarded, and slow learners are used in referring to them. In the school culture where reading is the task, they suffer ego deflation; they begin to feel inferior. Psychologically they retire from the competition and many are reluctant to try again even after they do achieve readiness.

Help for students who are threatened by the tyranny of the average is available in some schools. Year after year and decade after decade there have been some teachers who help their students succeed despite all the media-reported school failures. The keys to success are, and have been, teachers who recognize differences, who are concerned about individuals. They are teachers who respect students and who in turn demand respect and order. They care and they are demanding.

Some children have been taught (or have learned almost by themselves) to read by age four or five. They have already learned that books provide pleasure and adventure. Other children, encouraged by their parents to "keep out of the way" are significantly deprived of verbal contact with adults. Children need time with books, reading, and linguistic contact with adults in order to learn what reading can mean.

Adolescence

Changing Roles for Parents and Adolescents

One of the more difficult tasks of parents is that of changing from one role to another as children steadily become more mature. Unless parents do change, children are hampered in their attempts to grow up. For example, infants are totally dependent on care givers for food, clothing, and protection from disease and danger. Infancy quickly passes and children must learn to dress and exercise increasing self-care. However, they may be unable to demonstrate discretion about choice of clothes, where to play, and when to eat and sleep. Parental care and direction replace the single role of infant care. All too soon, as children grow, responsibility in increasing degrees must be transferred to adolescents. The direction given by parents must be more subtle than was the case before the teen years had been reached. Obviously, the changing role of parents is paralleled by role changes for teenagers.

The difficulty of shifting parental roles is intensified during the adolescent years by several factors:

1. Rapid change in social, moral, economic, family and technological phenomena cause adolescent and parental viewpoints to differ. Culture, as never before, is in a state of flux. Many adults, as well as adolescents, are confused about authority, responsibility, values, ethics, religion, health, family function, and personal rights.

2. The expanding perspective of adolescents in this changing cultural scene magnifies the contrast between the world as it really is now and what it was when parents were adolescents.

3. The urge of adolescents to grow up, to be independent, to be their own person, makes docile obedience to parental dictate virtually impossible. It is, some persons wisely say, almost essential that adolescents—regardless of how helpful and kind parents are—rebel against parents.

4. Growth in cognitive ability to the stage of logical thinking (not restricted to but certainly increasingly prominent in adolescence) makes it probable that there will be varied perspectives on many issues.

The least, and maybe all, that can be done about the generation gap is to see that it does not widen because of the absence of, or limitations in, communication. Unless parents can learn to talk with sons and daughters as though adolescents have some sense, some sane justification for their beliefs, adults may need help. Help is available and often joyfully welcomed by mothers and fathers after they have once tried using the advice, knowledge, and experience of trained communication facilitators.

Steps Toward Improved Communication

The initial step in parent-adolescent communication is to discuss the nature of the communication process. This is called metacommunication, or discusing the dynamics of interpersonal verbal transactions. Such questions as: What is listening? What is your attitude toward what you just heard? Do you agree with or believe what is said? What do you understand by nonverbal communication?

Observers can assist in analyzing the dynamics of a communication transaction: Was the tone of the speaker assertive, questioning, bombastic? Was the stance of the listener bored, attentive, or deeply involved? An outsider can be objective because he is not involved in the message or the outcome.

Feedback facilitates continuation of the communication. This feedback can be in the form of asking and answering questions. It can mean proposing an alternative or modified statement. It may simply be a reflection—trying to state in other words the same thing that one thinks was just said. Feedback may be nonverbal: the listener turns away, seems uninterested, or leans forward to get every word, touches the arm of the speaker to show sympathy or concern. The focal items are: Is my feedback critical or supportive? What kind of feedback will encourage continued communication?

Telling others what they did, said, or implied, inhibits genuine communication. "You just said...," "You are being defensive . . .," "That's just an excuse..." leads to silence or argument and tends to cause speakers to give up. The alternative to attributive declarations is to describe your feelings and what you thought was being said. Thus instead of "You just said..." the response may be "I understand your words to mean...," "My feeling is that you are being defensive..." or "My feeling is that there are reasons back of your statements."

Judgmental words inhibit personal level communication. The goal of parent-adolescent dialogue is to improve mutual understanding—not to give instructions. If the adolescent has reason to believe parents are being dictatorial, inflexible, unfeeling, the talk sessions will remain just talk. If parents state that certain specific words (or antecedent actions) are stupid, selfish, or illogical, the sessions degenerate into accusation versus counteraccusation. Instead of a judgmental stance, dialogue should be descriptive, "This is what I saw..." The obligation must be on the one who speaks, not on others.

Feelings are a legitimate concern in communication. It is asking too much of each other to expect a deep level description of feelings at initial sessions. The description of feelings might begin with impersonal things such as social events or physical environment. "This is the kind of neighborhood in which I live." It is a little more personal to say, "My home is warm, loving..." It is still more personal to say, "I feel in tune with the world..." Sometimes it is justifiable to say, "I'm really having a bad day today" (Harris & Howard, 1985).

Levels of Communicative Dialogue

The aim of parent-adolescent dialogue is to reach more tolerant acceptance and achieve more empathic understanding. Adolescents will come to appreciate their parents' reluctance to let them go, for parents to release them from the childhood status of dependence. The parents will come to see that their basic role is to assist adolescents to become independent rather than prolong childhood dependence. Both the release and the independence are important during this period. When adolescents and parents come to appreciate that the others are also facing difficulties they are in better positions to give mutual aid. An example will illustrate.

> The Link family came to the counseling center, ostensibly because Art, the eighth-grade son, was having trouble at school. Teachers could not give him enough attention—or the right kind. The principal recommended family counseling. The first three sessions were difficult but revealing Each family member blamed another for the problems. The son was merely the scapegoat. And Art was protesting at home and school. The father and mother had stopped talking to each other because they ended up screaming at one another about how to treat Irene, the high-

school senior daughter. She was skipping school and staying out late, sometimes overnight. Each person was encouraged to describe what was going on—without interruption from others. There was some head-shaking. This soon became a matter of discussing one's feelings. In the fourth session the father tearfully said, "I'll have to take the blame for all this mess. I have been having an affair with a woman I see when my sales route takes me to _____. We, as a family, have never discussed this; but it is apparent that all of us—includinding Art—know that something of the sort is going on. These sessions have made me realize that it's no secret. Actually, the woman-friend is not what I really want . . ."

Slowly, but eventually, each family member accepted some responsibility for the extent to which the supposed secret had distorted honest communication and influenced behavior.

Characteristics of Communication

The changing role of parents and adolescents daughters and sons might well be paraphrased into a definition of discussion. Dillon (1984) surveyed research on communication and reported the distinguishing features of discussion to be:

> Concerns, matters which have unknown and variable outcomes;
> Calls for opinions rather than facts;
> Outcomes are open-ended (i.e., no single conclusion is sought);
> Various points of view are solicited, and these views are listened to and weighed judiciously;
> Calls for thinking processes and respect for differences;
> Judgments, rather than hard data, are sought;
> Values, freedom, and reasonableness are essential;
> "Wait time" for responses must not be overruled by glibness;
> Respect for persons is pervasive.

These criteria and characteristics show that directing, drilling, training, and indoctrinating are not, realistically, discussion. The criteria show that respect for persons is essential. An observer of the counseling sessions with the Link family would have seen these distinguishing features of discussion in action. Most of the criteria features were developed during the sessions. Dillon (1984) emphasizes that participants' attitudes and predispositions are key factors in developing discussion. He perceives the processes of parents' and adolescents' adjusting to their changing roles as being the embodiment of a description of discussion.

EARLY AND MIDDLE ADULTHOOD

Communication, Friendship, and Therapy

Parents counsel with children as the children grow. Parents team with teachers to counsel elementary students. Teachers, counselors, and parents work to help adolescents resolve their conflicts and developmental challenges. College age students typically have college counselors. Such ready availability of help ceases abruptly for young adults. They find themselves "on their own" when they pursue adjustment in the world of work, marriage, and strange communities. They have moved away from parents and out of the school setting. This also removes them from their familiar milieu of family, friends, and frequently seen teachers. The opportunity to receive support, advice, and to obtain release from tension through dialogue has almost disappeared.

The authors believe that if there were a most difficult age-stage of life it would be the lonely time of early adulthood. The incidence of emotional breakdown by age groups attests to the validity of this belief. Persons, from childhood on, need friends, confidants, and counselors. In fact, counseling has been called the purchase of friendship. Major developmental thrusts that may derive from friendship—purchased or gratuitous—are:

> Release or partial release from bottled-up emotion; i.e., they receive emotional catharsis.
>
> Altered perspectives on personal problems—this may occur just from the audible description of dilemmas, with no advice from communicating partners.

Perception of alternative routes of behavior stemming from partner's advice, suggestion, or personal reflection.

Fortunately, there is some increase in the availability of formal counseling and in the inclination to seek assistance. There are growing numbers of therapeutic groups in community clinics, churches, and private practice. The Alcoholics Anonymous approach had encouraged the growth of counseling groups for compulsive gamblers, child abusers, drug abuse, sexual deviation, parents without partners, and dialogue on death and dying. Some employers have used human relations experts to lead groups in matters of supervisor-worker relationships, personal problems of employees, and in-plant policy.

Unfortunately, many persons have the erroneous notion—in our independent, self-sufficient, self-made man society, and the tradition of the robust American—that to seek help is to reveal weakness. The truth is that seeking help in time of trouble is a strength. It takes courage to tell a friend, spouse, counselor, or personal development group about one's failures, inadequacies, and selfish thoughts. It takes courage to decide that one will take charge of one's life rather than just to continue to drift and endure. Currently there is mounting interest in group processes (McGrath & Kravitz, 1982).

Marriage and Communication

Because of the frequency of divorce, separation, and desertion many studies have sought to determine the causes of marital discord. The more frequently cited disrupting factors are: difference of age, socioeconomic class, religion, educational level, and ethnic origins. Low income, emotional immaturity, divorce in parental family, wife's working, spending habits, and sexual incompatibility have also been cited as sources of potential disruption. Because there have been successful marriages where the above factors seem to be contradictory, better explanations have been sought. The factor which is currently emerging as most prominent in marriage failure, or success, is the matter of communication.

The Family Service Association of America polled thousands of couples and found that 87 percent of them said their main problem was poor communication. If marriage partners can learn to listen—and the art can be taught and learned—they can become sensitive to the needs, thoughts, and wishes of their spouses. Rook (1984) concludes that listening is really a form of caring. The readers can check it for themselves: Watch a mother as she listens to her preschooler tell about his day in kindergarten. Watch a young couple in a restaurant. Watch a teacher with a sad or lonely student.

Parents Without Partners is a rapidly growing organization which provides opportunities for otherwise isolated parents to meet others of those 20 million parents who have no spouse. Because of death, desertion, or divorce thousands of parents have to try to fill two roles. When the need such people have for communication is unmet, their loneliness triggers depression. Depressed persons are frequently those who have minimized, never developed, or lost their communication with others. Because of this phenomenon it is easy to understand the growing number of groups that seek to establish communication with others—especially those suffering similar burdens: divorce, parents without partners, alcoholics, families of alcoholics, parents of handicapped children, families of the terminally ill, and parents of gay sons or daughters.

At the University of Minnesota Olson and his colleagues have focussed on preparation for marriage as a form of divorce prevention. While this emphasis is fairly common, the methods at Minnesota are unique. An instrument was devised to assess the relationship of engaged couples and indicate probability of marital success. PREPARE, the premarital inventory, consists of 125 items designed to identify strengths and work areas in eleven relationship areas: communication, realistic expectations, personality issues, conflict resolution, financial management, leisure activities, sexual relationship, children and marriage, family and friends, equalitarian roles and religious orientation. These scales were constructed in such a way as to promote couple dialogue and to enable mutual reflection.

In order to determine the predictive validity of PREPARE, a three-year followup study was conducted with 164 couples who had been administered in the instrument during their engagement. It was found that the scores obtained three months before marriage could forecast with 90 percent accuracy which couples would separate and divorce as well as those who would develop a satisfying marriage. It seems clear that PREPARE can be helpful in identifying those engaged couples who are at risk in terms of eventual divorce. Some of the options couples may wish to consider are: lengthen their engagement for further preparation; seek professional guidance in whatever areas (e.g. communication, conflict resolution, personality issues or role relationships) where their scores suggest incompatibility; and/or reconsider the plans for marriage. About 10 percent of the participants cancelled their marriage plans after completing the two standard feedback sessions involving both parties (Fowers and Olson, 1986).

The Art of Listening

Psychological consultants, for decades, have recognized the importance of listening as a factor in facilitating therapeutic transactions. Effective teachers, intuitively it seems, also have recognized the importance of listening in building teacher-pupil rapport and classroom morale. Business executives, increasingly, are using lecturers, seminars, and workshops on listening to facilitate sales and services. The American Telephone and Telegraph uses the concept in its advertising slogan, "Reach out and touch someone." Empirical evidence proves that persons can become better listeners by purposefully focussing their attention on the speaker's word, face, and gestures.

A first step toward better listening is to admit that it is more a function of what you are rather than what you do. One's concern must be on the other person: e.g., the teacher-listener must be a loving, or at least accepting and empathic person. The point is illustrated by the mother of an infant. She watches and becomes thrilled by the sounds of the infant. She can hear in those attempted vocalizations words and communication. Concerned teachers get similarly excited when they hear and share students' talk about their victories, adventures, and disappointments.

Children are quick to learn when they are being listened to. A twelve-year-old boy who had lived in several unhappy foster homes said to his youth counselor, "You're not listening to me. You're just thinking about what you're going to do next." Most children are not so blunt or articulate. They simply turn off the good advice of the nonlistener like they do when the advertisements come on television—which also provides no feedback. The turn-off button is a mental one. Teachers who fail to listen are tuned out. Parents who repetitiously complain and accuse are tuned out. Most of us tune out the one-way speakers just as we tune out the traffic noise on a busy street.

A middle-aged man recalled an instance from his college days in which taking the time to listen mattered. One of his roommates was a person "I just didn't like, a guy who had no friends, no dates. One evening after a basketball game I walked back to the dorm with him and tried to show friendship by stopping for a milkshake. He told me about his seriously ill father and life of poverty. Years later I learned that the man had decided, just before the incident, to quit school. That night he changed his mind, was graduated, and went on to do well in his profession."

An anecdote of a first-grade boy also illustrates that a moment may greatly transcend the time it really occupies. The boy and his mother had stopped for lunch at a restaurant. After taking the mother's order the waitress asked the boy what he wanted. His mother answered for him, "Plain hamburger on rye bun." The waitress again turned to the boy and asked if he wanted lettuce or relish. The mother said, "Lettuce, no relish." The waitress tried again, "Do you want catsup or mustard?" Mother said, "A little catsup." As the waitress left the table she said to the boy, "My son is just about your age. He likes hamburgers just like you are having." The boy turned to his mother and said, "Gee, Mom, she thinks I'm real."

Developmental Groups

Neither individual psychological therapy nor developmental groups will resolve all the problems of communication that adults encounter. It is, however, estimated that as many as 80 percent of participants gain from involvement. Such a figure must be highly approximate because some persons are transformed, others are helped somewhat, certain parties are unaffected, and still others may be harmed. The success is, in part, determined by expectations. A group experience may be successful because one has come to know which of one's actions, tones, gestures are influencing others positively or negatively. It may be unsuccessful because nothing can be done to effect change; e.g., one's gravelly voice, protruding chin. Under trained leaders, participants are typically helped with personal problems, marital discord, and work difficulties (McGrath & Kravitz, 1982).

Individual therapy has declined in popularity in recent years. This is partly because of expense; plus the fact that ultimately success depends on returning to group milieus. Group processes are becoming more reliable. Our concern will focus on group processes led by trained, qualified, and certified facilitators.

"Let the buyer beware" is an advisable approach to participation in a group. There are quacks (e.g., the sex specialists who volunteer, or demand intercourse as a part of therapy); there are those who rely on supposedly inspired admonitions because they lack training, experience, and empathic insight to the complexity of human relationships. There are also group facilitators who are well trained, empathic, concerned with professional ethics, and who have served internships. Typically, one can begin to evaluate the qualifications of the prospective facilitator by virtue of affiliation with a medical center or university. But if those with such affiliations make promises as to what the personal outcome will be, the "caution" sign is out. This is because in some groups there will be persons who, for reasons of past history, attitudes, or specific expectations, do no profit. You might be that person—without any reflection on your worth. Possibilities (but no promises) should be expressed by the prospective facilitator.

Choosing a facilitator can begin by communicating "What are your qualifications and experience?" Carey (1985) suggests that the client interview the counselor. The chooser should put some reliance on his impressions of the counselor as a person. Is he warm, responsive, attentive, or likeable? After all, if the group is to be productive, the client is going to be sharing some thoughts and secrets that are usually considered to be private. There is almost a certainty of dealing with sensitive areas of personality. However, for this very reason, privacy, the client cannot rely completely on first impressions. The fact that individuals do have critical problems or acknowledged personality difficulties may tend to render them unduly suspicious. Hence, the reputation of the counselor should not be disregarded because of a negative first impression. The would-be clients might ask themselves why certain acts or appearances of the counselor create the personal impressions they do. The answer may reside more in the client than in the counselor.

LATER ADULTHOOD

Maintaining and Reforming Interpersonal Contacts

Current research focusses on three ways in which communication and social relationships support healthy development. (1) Humans are inherently social creatures and they need support during times of crises. The support of others decreases vulnerability to stress related disorders. (2) The phenomenon of loneliness emphasizes the basic need for human ties. Loneliness indicates that the absence of social ties creates emotional distress. (3) Social isolation creates a predisposition to deviant behavior. Communication and social relationship provide the support which are an antidote to suicide, alcoholism, and drug abuse. The incidence of all these tragedies is high among the elderly (Rook, 1984).

Loneliness, social isolation, and the need for social support are not confined to the older years; but, the incidence is high among the elderly and these three items merit specific mention. Ways in which social support can be given are rather obvious when children and adolescents are discussed. Support for the elderly is less obvious because the extent of the problem is of recent origin and because of the reluctance of the elderly to admit dependence. Social isolation is so common among the elderly that "disengagement" is cited as one theory of aging. Interaction with the environment (which is reduced in later adulthood) is essential in lifespan development.

Family Ties and Individual Perspective

The personal impact of retirement is variable. Whether or not retirement is gratifying depends on perspective. Some persons are grateful for additional time with spouse; others say they need relief from the constant presence of one's mate. A study of Sun City, Arizona—a unique city because it restricts home ownership to older persons (thus few children and no school taxes)—indicates that some people like it because it affords maximum contact with age peers. There is something for everyone because the recreation centers provide clubs, companionship, and communication with persons with like interests, hobbies, and developmental pursuits. If there is no class or club available in a specific interest area, one can be readily organized. Sun City provides relief from the chaos and chatter of young children. Simultaneously it provides boredom, loneliness, and isolation for men and women who love the vigor and unpredictability of the young. Some persons choose to live in Sun City because they can finally sever the binding ties with grown sons and daughters. Others have sold their Sun City homes and returned to their lifelong hometowns because their communication with sons and daughters was cut off or abbreviated.

Gober and Zonn (1983) found that communication with family and friends was the major element in determining migration to Sun City or staying at home; or having moved to Sun City, remain a migrant or return home. "The warm, dry climate of Arizona was clearly an attraction to migrants" (p. 292) but; whether Sun City lifestyle and weakened communication with relatives and former friends is gratifying depends on individual responses. This investigation indicates the need for the elderly to communicate with other people in similar situations. It is both an opportunity to anticipate personal responses and to adapt attitudes when it is deemed desirable. Seminars, discussion groups, and support sessions may act in retirement as do communication groups when dealing with alcohol, drug abuse, single persons, terminally ill, or parents of handicapped. Whatever the age, communication facilitates adaptation to change.

Arranging Intergenerational Communication

Estoya Whitley, a second-grade teacher at the University of Florida's laboratory school, wanted to develop an Adopted Grandparents Program for her students and the residents of nearby Hillhaven Nursing Home. That was 20 years ago. When the concept was proposed to the elderly patients, they were reluctant to be adopted feeling that "the children will probably visit a couple of times and never come back just like everybody else." On the contrary, this program has continued with six to eight children visiting the nursing home on a daily basis. Occasionally some of the residents come to school in wheelchairs guided by the students. During the time they spend together the emphasis is on communication through visiting, writing letters, reading, singing, playing games, drawing, and making things.

The greatest benefit for children is growth in their level of maturity. They develop compassion for the handicapped, a commitment to helping older friends, learn to assist grandparents in coping with infirmities, acquire a positive attitude toward aging and gain valuable insights about death and dying. When one of the elderly participants dies, the loss must be dealt with so that the children can work through and understand their grief. To illustrate, on each visit during the weeks preceding her death, Grandma Ruth gave the children one of her cherished possessions. She said, "Russell, you take this African violet to remember me by." "Vickey, I want you to plant these flower seeds and remember how I told you to care for them." During this period Ruth also returned the gifts that the children had made for her. In every case she told them how much she appreciated their kindness. Later, when she died, the children's first reaction was silence. Then their expression of feelings about losing a friend was encouraged by the teacher. Grief should be expressed and shared. The discussions focus on good memories and reflection. "She was really a nice Grandma to us." "She said we brought a lot of sunshine into her life because we were good to her." "She lived a long time." "Everybody has to die." After expressing their feelings the children send messages of sympathy to surviving family members (Whitley & Duncan, 1985).

According to the Hillhaven staff, the patients who have children to communicate with on a regular basis show a higher level of energy, take more pride in their personal appearance, develop improved appetites, make fewer complaints, and look forward to the visits. The children are often as helpful as adults in teaching stroke victims to speak and the old people try harder for them. The nursing director says, "These boys and girls can get the grandparents to do things no adult could possibly get them to do. The teacher and her students fill the nursing home with a love, excitement, and happiness that brings life to many otherwise lonely and lifeless elderly patients" (Whitley & Duncan, 1985).

A major factor in successful aging is learning new skills of communication. It is important to continue sharing feelings and ideas with spouses, to interact with sons and daughters, to converse with agemates who face similar problems and challenges, and to listen to children whose optimism encourages a healthy view of life and the future.

REFERENCES

Achenback, T. M., & Edelbrock, C. S. (1984). Psychopathology of childhood. *Annual Review of Psychology, 35,* 227–256.

Aiken, L. R. (1982). *Later Life.* New York: Holt, Rinehart & Winston.

Allcorn, S. (1985). The knowledge gap of adult education. *Lifelong Learning, 8*(5), 12–16.

Allman, W. (1985, October). We have nothing to fear. *Science, 6*(8), 38–41.

American Medical Association (1983). *Educating children for the new era of aging.* Monroe, WI: author, 1–4.

Anderson, L. (1983). *The aggressive child.* Washington, DC: U.S. Department of Health and Human Services.

Anderson, R. Kinney, J., & Gerler, E. (1984). The effects of divorce on children's classroom behavior and attitudes toward divorce. *Elementary School Guidance and Counseling, 19*(1), 70–76.

Apgar, V., & Beck, J. (1972). *Is my baby all right?* New York: Simon & Schuster.

Ardrey, R. (1966). *The territorial imperative.* New York: Dell Publishing Company.

Ardrey, R. (1970). *The social contract.* New York: Dell Publishing Company.

Arlin, P. (1977). Piagetian operations in problem finding. *Developmental Psychology, 13,* 297–298.

Averill, J. N. (1984). The acquisiton of emotions during adulthood. In C.Z. Malatesta & C. E. Izard (Eds.), *Emotions in adult development* (pp. 23–43). Beverly Hills, CA: Sage Publications.

Baldwin, C., Colangelo, N., & Pittman, D. (1984). Perspectives of creativity throughout the lifespan. *The Creative Child and Adult Quarterly, 9*(1), 9–17.

Baltes, P., & Schaie, K. W. (1974). Aging and IQ: The myth of the twilight years. *Psychology Today, 7*(1), 35–40.

Bandura, A. (1977). *Social learning theory.* Englewood Cliffs, NJ: Prentice-Hall.

Bardwick, J. M. (1978). Middle age and a sense of future. *Merrill-Palmer Quarterly, 24* 129–138.

Bayley, N. (1949). Consistency and variability in the growth of intelligence from birth to eighteen years. *Journal of Genetic Psychology, 75,* 165–196.

Benedict, R. (1935). *Patterns of culture.* New York: Mentor Books.

Bengston, V., & Robertson, J. (1985). *Grandparenthood.* Beverly Hills, CA: Sage Publications.

Bennett, W. (1985, August). New help for the deaf. *The Harvard Medical School Letter, 10*(9), 4, 5.

Bennett, W. (1985a, July). The last gasp? *The Harvard Medical School Health Letter, 10*(1), 6.

Benson, H., & Proctor, W. (1984). *Beyond the relaxation response.* New York: New York Times Books.

Berger, R. (1983). Health care for lesbians and gays. *Journal of Social Work and Human Sexuality, 1*(3), 54–73.

Berk, S. F. (1980). *Women and household labor.* Beverly Hills: Sage Publications.

Bezold, C. (1985, August). Drugs and health in the year 2000. *The Futurist, 19*(4), 36–39.

Binet, A., & Simon, T. (1905). Methodes nouvelles pour le diagnosis da niveau intellectual des anormaux. *L'Annee Psychologique, 11*(9), 244.

Birren, J. (1970). Toward an experimental psychology of aging. *American Psychologist, 25,* 124–135.

Bonner, H. (1965). *On being mindful of man.* Boston: Houghton Mifflin.

Bower, T. G. R. (1982). *Development in infancy.* San Francisco: W. H. Freeman.

Bowlby, J. (1980). *Attachment and loss.* New York: Basic Books.

Bradley, R. H., & Teeter, T. A. (1977). Perceptions of control over social outcomes and student behaviors. *Psychology in the Schools, 14,* 230–235.

Bradshaw, J., & Nettleton, N. (1983). *Human cerebral asymmetry.* New York: Prentice-Hall.

Brazelton, T. B. (1983). *Infants and mothers.* New York: Delacorate Press.

Brent, D. (1985, October 24). *Drugs and teenage suicide.* Paper presented at the American Academy of Child Psychiatry, San Antonio, Texas.

Brewer, G., & Brewer, T. (1985). *What every pregnant woman should know.* New York: Viking Penguin.

Bronfenbrenner, U. (1984, Spring). The changing family in a changing world. *Peabody Journal of Education, 61*(3), 52–70.

Brubaker, T. (1985). *Later life families.* Beverly Hills, CA: Sage.

Bruner, J. (1982, January). Schooling children in a nasty climate. *Psychology Today, 16*(1), 57–63.

Bruner, J. (1983). *In search of mind.* New York: Harper & Row Publishers.

Buhler, C. (1968). The course of life as a psychological problem. *Human Development, 11,* 184–200.

Bumpass, L. (1984). Children and marital disruption: A replication and update. *Demography, 21,* 71–82.

Burish, T., Levy, S., & Mayerowitz, B. (1985). *Cancer, nutrition, and eating behavior.* Hillsdale, NJ: Lawrence Erlbaum Associates.

Butler, R., & Gleason, H. (1985). *Productive aging.* New York: Springer Publishing Company.

Cagle, M. (1985). A general abstract model of creative thinking. *Journal of Creative Behavior, 19*(2), 104–109.

Cannon, W. (1915). *Bodily changes in pain, hunger, fear, and rage.* New York: Appleton-Century-Crofts, Inc.

Caplan, F. (1982). *The first twelve months of life.* New York: Putnam Publishing.

Carey, J. (1985). Bleak days for psychiatry—A search for answers. *U.S. News & World Report, 98*(7), 73–74.

Carnegie Corporation (1985, Fall/Winter). Renegotiating society's contract with the public schools. *Carnegie Quarterly, 29*(4), 1–12.

Chance, P. (1986). *Thinking in the classroom.* New York: Teachers College Press.

Chandler, T. A., Wolf, F. M., Cook, B., & Dugovics, D. A. (1980). Parental correlates of locus of control in fifth graders. *Merrill-Palmer Quarterly, 26,* 183–195.

Children's Defense Fund (1985). *Black and white children in America.* Washington, DC: Author.

Chomsky, N. (1972). *Language and the mind.* New York: Harcourt, Brace & Jovanovich, Inc.

Cloward, R. (1976). Teenagers as tutors of academically low-achieving children. In V. Allen (Ed.), *Children as teachers: Theory and research in tutoring* (pp. 219–230). New York: Academic Press.

Cohen, P., Kulik, J., & Kulik, C. (1984). Educational outcomes of tutoring:

A meta-analysis of findings. *American Educational Research Journal,* 237–248.

Comfort, A. (1981, December). Avoiding shortevity. *Psychology Today, 15*(12), pp. 110–112.

Cousins, N. (1984). What you believe and feel can have an effect on your health. *U.S. News & World Report, 96*(3), 61–62.

Cox, J., Daniel, N., & Boston, B. (1985). *Educating able learners: Programs and promising practices.* Austin: University of Texas Press.

Darnell, J. (1985, October). RNA. *Scientific American, 253*(4), 68–87.

DeBono, E. (1984). *Tactics: The art and science of success.* Boston: Little Brown & Co.

Dentzer, S. (1985, May 6). Has Sun City come of age? *Newsweek, 105*(18), p. 68.

Dillon, J. T. (1984). Research on questioning and discussion. *Educational Leadership, 42*(3), 50-56.

Donaldson, M. (1978). *Children's minds.* New York: W. W. Norton.

Driscoll, F. (1972). TM as a secondary school subject. *Phi Delta Kappan, 54*(4), 236–237.

Dunn, R. (1984). Learning style. State of the science. *Theory Into Practice, 23*(1), 10–17.

Durkheim, E. (1952). *Suicide.* London: Routledge & Kegan Paul, pp. 2–48.

Eichhorn, D. H. (1980). The school. In the 79th Yearbook of the National Society for the Study of Education, Part I, M. Johnson (Ed.), *Toward adolescence: The middle years* (pp. 56–73). Chicago: University of Chicago Press.

Eisenberg, A., & Eisenberg, H. (1985), July/August). Midlife without crisis. *American Health, 4*(6), 60–69.

Elderhostel (1986). *This is Elderhostel.* Boston: Author.

Elkin, F., & Handel, G. (1984). *The child and society: The process of socialization.* New York: Random House.

Elliott, D., Huizinga, D., & Ageton, S. (1985). *Explaining delinquency and drug use.* Beverly Hills, CA: Sage Publications.

Ennis, T. (1985). *Alzheimer's disease and related disorder Association Newsletter, 5*(3), 1–12.

Epstein, H. T. (1978). Growth spurts during brain development: Implications for educational policy. In National Society for the Study of Education, 77th Yearbook, Part II, J. Chall & A. Mirshy (Eds.), *Education and the Brain* (pp. 343–370). Chicago: University of Chicago Press.

Erikson, E. (1963). *Childhood and society.* New York: W. W. Norton.

Erikson, E. (1980). *Identity and the life cycle.* New York: W. W. Norton.

Eron, L. D., & Peterson, R. A. (1982). Abnormal behavior: Social approaches. *Annual Review of Psychology, 33,* 231–264.

Eurich, N. (1985). *Corporate classrooms: The learning business.* New York: Carnegie Foundation for the Advancement of Teaching.

Euster, G. (1982, Fall). Serving older adults through institutions of higher education. *Gerontology and Geriatrics, 3*(1), 69–75.

Farran, D. (1982, September). Now for the bad news. *Parents, 57*(9), 80–83.

Feldman, H., & Feldman, M. (1985). *Current controversies in marriage and family.* Beverly Hills, CA: Sage Publications.

Felsenfeld, G. (1985, October). DNA. *Scientific American, 253*(4), 58–67.

Fogg, R. (1985, June). Creative peaceful approaches for dealing with conflict. *Journal of Conflict Resolution, 29*(2), 330–358.

Fowers, B. J., & Olson, D. H. (1986). Predicting marital success with PREPARE: A predictive validity study. *Journal of Marital and Family Therapy.*

Fraiberg, S. (1967, December). The origins of human bonds. *Commentary, 44,*(6), 51–57.

Freud, S. (1925). *On creativity and the unconscious.* New York: Harper & Row.

Freud, S. (1935). *An autobiographical study.* London: Hogarth Press.

Freud, S. (1962). *The ego and the id.* New York: W. W. Norton.

Friedrich, O. (1985). What do babies know? In H. Fitzgerald & M. Walraven (Eds.), *Human development* (pp. 83–88). Guilford: Dushkin Publishing Group.

Frymier, J. (1984). *One hundred good schools.* West Lafayette, IN: Kappa Delta Pi.

Gagne, E. (1977). *The cognitive psychology of school learning.* Boston: Little, Brown & Company.

Gagne, R. (1977). *The conditions of learning.* New York: Holt Rinehart & Winston.

Gale, A. (1985). *Handbook on aging for prekindergarten through grade 6.* Chicago: Department on Aging and Disability.

Galloway, C. (1976). *Psychology of learning and teaching.* New York: McGraw-Hill Book Company.

Gallup, G. (1985). Forecast for America. *Television and Families, 8*(1), 11–17.

Gallup, G. (1985a). *Youth survey on marriage and divorce.* Princeton: Gallup Research Center.

Gardner, H. (1983). *Frames of mind: The theory of multiple intelligences.* New York: Basic Books.

Gardner, H. (1985). The varieties of intelligence. *New Ideas Psychology, 3*(1), 47–65.

Garner, D., & Garfinkel, P. (1985). *Handbook of psychotherapy for anorexia ner-*

vosa and bulmina. London: Guilford Press.

Gaylin, W. (1984). *The rage within: Anger in modern life*. New York: Simon & Schuster.

Gelles, R., & Cornell, C. (1985). *Intimate violence in families*. Beverly Hills, CA: Sage.

Gelman, D. (1985, May 9). Who's taking care of our parents? *Newsweek*, pp. 61–68.

Ghiselin, B. (1952). *The creative process*. New York: Mentor Books.

Glass, B. (1969). Evolution in human hands. *Phi Delta Kappan, 50,* 506–510.

Glieberman, H. A. (1981, July 20). Who so many marriages fail. *U.S. News & World Report, 91*(3), 53–55.

Gober, P., & Zonn, L. (1983). Kin and elderly amenity migration. *The Gerontologist, 23*(2), 288–294.

Goldman, R., & Goldman, J. (1982). *Children's sexual thinking*. London: Routledge & Kegan Paul.

Goleman, D. (1984, May). The faith factor. *American Health, 3*(3), 48–53.

Goleman, D. (1985). *Vital lies, simple truths: The psychology of self-deception*. New York: Simon & Schuster.

Goleman, D. (1986). What's your stress style? *American Health, 5*(3), 41–45.

Green, R., & Kolevzon, M. (1984, October). Characteristics of healthy families. *Elementary School Guidance and Counseling, 19*(1), 9–18.

Guilford, J. (1977). *Way beyond the IQ*. Buffalo, NY: Creative Education Foundation.

Guilford, J. (1979). Some incubated thoughts on incubation. *Journal of Creative Behavior, 13*(1), 1–8.

Guilford, J. (1981). *The structure of intelligence*. New York: McGraw-Hill Book Company.

Hall, G. S. (1904). *Adolescence: Its psychology and its relations to physiology, anthropology, sociology, sex, crime, religion, and education*. New York: Appleton.

Halverson, C. F., Jr., & Victor, J. B. (1976). Minor physical anomalies and problem behavior in elementary school children. *Child Development, 47,* 281–286.

Harper, P. (1985, August 18). Gene mapping and medical genetics. *Medical Genetics, 22*(4), 241-242.

Harrington, B., & Harrington, J. (1985). Can teachers spot a bright child? *The Creative Child and Adult Quarterly, 10*(1), 34–36.

Harris, G., & Gurin, J. (1985, March). Look who's getting it all together. *American Health, 4*(2), 42–47.

Harris, I., & Howard, K. (1984). Parental criticism and the adolescent experience. *Journal of Youth and Adolescents, 13*(2), 113–121.

Havighurst, R. (1972). *Developmental tasks and education.* New York: David McKay.

Hawley, W. Rosenholtz, S., Goodstein, H., & Hasselbring, T. (1984, Summer). Good schools: What research says about improving student achievement. *Peabody Journal of Education, 61*(4), 15–52.

Helms, D., & Turner, J. (1986). *Exploring child behavior.* Monterey: Brooks/Cole.

Hemingway, E. (1965). *A moveable feast.* New York: Charles Scribner.

Hillson, H., & Myers, F. (1963). *The demonstration guidance project 1957–1962.* New York: George Washington High School Board of Education.

Hofmeister, A. M. (1984). The special educator in the information age. *Peabody Journal of Education, 62*(1), 5–21.

Holmes, K., Quinn, T., Corey, L., & Cates, W. (1985). Sexually transmitted disease. *Urban Health, 14*(3), 29–35.

Howard, T., & Rifkin, J. (1978). *Who should play God?* New York: Dell Publishing Company.

Howell, M. (1985). Is daycare hazardous to health? In J. McKee (Ed.), *Early childhood education* (pp. 99–100). Guilford: Duskin Publishing Group.

Huessman, L., Eron, L., Lefkowitz, M., & Waldor, L. (1984). Stability of aggression over time and generations. *Developmental Psychology, 20*(6), 1120–1134.

Hunt, S. L. (1983 Winter). Stress without distress. *Kappa Delta Pi Record, 19*(2), 38–41.

Isakson, S., & Parnes, S. (1985). Curriculum planning for creative thinking and problem solving. *Journal of Creative Behavior, 19*(1), 1–29.

James, W. (1899). *Talks to teachers.* New York: Holt & Company.

Jenkins, C. (1985). New horizons for psychosomatic medicine. *Psychosomatic Medicine, 47*(1), 3–25.

Jensen, A. (1969). How much can we boost IQ and scholastic achievement? *Harvard Educational Review, 39,* 1–123.

Jensen, A. (1980). *Bias in mental testing.* New York: Free Press.

Jensen, A. (1984, November). The limited plasticity of human intelligence. *New Horizons, 25,* 18–22.

Jersild, A. (1968). *Child psychology.* Englewood Cliffs, NJ: Prentice-Hall.

Juhasz, A. (1983). Early adolescent preceptions about the need for adults to know more about them. *Journal of Early Adolescence, 3*(4), 305–315.

Kastenbaum, R. (1986). *Death, society, and human experience.* St. Louis, MO: C. V. Mosby Co.

Katz, L. G. (1977). What is basic for young children? *Childhood Education, 54*(1), 16–19.

Kelly, G. (1980). *Sexuality: The human perspective.* Woodbury, NY: Barrons'

Educational Series.

Kerman, S. (1979). Teacher expectations and student achievement. *Phi Delta Kappan, 60,* 716–718.

Kilmann, R. H. (1985). Corporate culture. *Psychology Today, 18*(4): 62–68.

Kim, J. (1985) Effects of nutrition on disease and lifespan. *The Journal of Applied Nutrition, 37*(1), 41–42.

Klaus, M., & Kennell, J. (1983). *Bonding: The beginnings of parent-infant attachment.* New York: New American Library.

Kobasa, S. (1984, September). How much stress can you survive? *American Health,* pp. 64–77.

Koffka, K. (1935). *Principles of Gestalt psycholgy.* NY: Harcourt, Brace & Co.

Kohlberg, L. (1983). *The psychology of moral development, Vol. II.* San Francisco: Harper & Row.

Kohler, W. (1925). *The mentality of apes.* New York: Harcourt, Brace & Co.

Kolata, G. (1985, September). What causes nearsightedness? *Science, 229,* (4719), 1208–1213.

Kornhaber, A. (1986). *Between parents and grandparents.* New York: St. Martin's Press.

Krech, D. (1969, March). Psychoneurobiochemeducation. *Phi Delta Kappan, 50*(7), 370–375.

Kubler-Ross, E. (1969). *On death and dying.* New York: Macmillian Publishing Co.

Kubler-Ross, E. (1979) *Death: The final stage of growth.* Englewood Cliffs, NJ: Prentice-Hall.

Lamaze, F. (1984). *Painless childbirth.* Chicago: Contemporary Books.

Lauer, J., & Lauer, R. (1985, June). Marriage made to last. *Psychology Today, 19*(6), 22–26.

LeBow, M. (1983). *Child obesity.* New York: Springer Publishing Co.

Lehman, H. (1966, April). The psychologist's most creative years. *American Psychologist, 21*(4), 363–369.

Leiblum, S., & Pervin, L. (1985). *Principles and practices of sex therapy.* London: Guilford Press.

Leich, J., & Pieper, H. (1982, Fall). Education: A meaningful leisure alternative for the elderly. *Activities, Adaptation and Aging, 3*(1), 37–45.

Levinson, D. (1978). *The seasons of man's life.* New York: Alfred A. Knopf.

Lewin, K. (1936). *Principles of topological psychology.* New York: McGraw-Hill Book Company.

Lewin, K. (1948). *Resolving social conflicts.* New York: Harper & Row.

Lewis, M. L., & Butler, R. N. (1985, February-March). The factss of later life. *Modern Maturity, 28*(1), 59–60.

Linden, F. (1985, July 15). Yuppies have appeal but older Americans have assets. *Business Weekly,* 2903, p. 24.

Lindsley, D. B. (1951). Emotion. In S. S. Stevens (Ed.), *Handbook of experimental psychology* (pp. 473–516). New York: John Wiley & Sons.

Lippitt, P. (1975). *Students teach students.* Bloomington, IN: Phi Delta Kappan Foundation.

Lipsitz, J. S. (1980). The age group. In 79th Yearbook of The National Society for the Study of Education, Part I, M. Johnson (Ed.), *Toward adolescence: The middle school years* (pp. 7–31). Chicago: University of Chicago Press.

Long, L., & Long, T. (1983). *The handbook for latchkey children and their parents.* New York: Arbor House.

Lowes, J. (1927). *The road to Tanadu: A study on the ways of imagination.* Boston: Houghton Mifflin.

Lynch, G., McGaugh, J., & Weinberg, N. (1984). *Neurobiology of learning and memory.* London: Guilford Press.

Lyon, H. (1981) Our most neglected natural resource. *Today's Education,* 70(1), 15–20.

Lytle, V. (1985, June). NEA's nine principles for educational revolution. *NEA Today,* 3(8) 4–5.

Maccoby, E. (1980). *Social development: Psychological growth and the parent-child relationship.* New York: Harcourt, Brace & Jovanovich.

Maddock, J. Neubeck, G., & Sussman, M. (1984). *Human sexuality and the family.* New York: The Haworth Press.

Malatesta, C., & Izard, C. (1984). *Emotions in adult development.* Beverly Hills, CA: Sage Publications.

Malecki, M. (1985, July/August). Working with families who donate organs and tissues. *Children Today,* 14(4), 26–29.

Maslow, A. (1962). *Toward a psychology of being.* Princeton, NJ: D. Van Nostrand Co.

Maslow, A. (1970). *Motivation and personality.* New York: Harper & Row.

Mayer, J. (1975). Obesity during childhood. In M. Winick (Ed.), *Childhood obesity* (pp. 73–80). New York: John Wiley & Sons.

McCabe, E. (1985, August 1). Creativity. *Vital Speeches of the Day, 51* (2), 628–632.

McGrath, J. E., & Kravitz, D. A. (1982). Group research. *Annual Review of Psychology, 3,* 195–230.

McMillan, P. J. (1982, July 12). An assassin's portrait. *New Republic,* 16–18.

Mead, M. (1928). *Coming of age in Samoa.* New York: William Morrow & Company.

Meddis, S. (1985, November 14). Missing children effort helps find 2,600. *USA Today*, p. 3A.
Meir, J. (1985). *Assault against children.* San Diego: College Hill Press.
Menninger, K. A., Mayman, M., & Pruyser. P. (1963). *The vital balance.* New York: Viking Press.
Miller, M. (1979). *Suicide after sixty.* New York: Springer Publishing Co.
Moos, R. (1985). *Coping with life crises.* New York: Plenum Pub. Co.
Murphy, M., & Donovan, S. (1983). A bibliography of meditation theory and research 1931–1983. *Journal of Transpersonal Psychology, 15*(2), 181–227.
Mussen, P. (1984). *Handbook of child psychology.* New York: John Wiley & Sons.
Naisbitt, J. (1982). *Megatrends.* New York: Warner Books.
Naisbitt, J. (1985, August). Megachoices: Options for tomorrow's world. *The Futurist, 19*(4), 13–16.
National Center for Research in Vocational Education (1986). *Career information in the classroom.* Bloomington, IL: Meredian Education Corporation.
National Institute on Aging (1982). *Special report on aging.* Washington, DC: U.S. Department of Health and Human Services.
Neugarten, B. L. (1980, February). Must everything be a midlife crisis? *Prime Time*, pp. 1–3.
Neugarten, B. L., & Hagestad, G. (1976). Age and the life course. In R. H. Binstock & E. Shanos (Eds.), *Handbook of aging and the social sciences.* New York: Van Nostrand & Reinhold.
Newsweek (1984, December 3). A slow death of the mind, pp. 56–62.
Nichols, M. (1986). *Turning forty in the eighties.* New York: W. W. Norton.
Nicolson, J. (1980). *Seven ages.* Glasgow: William Collins & Company.
Nicolayson, M. (1981). Dominion in children's play. In R. Strom (Ed.), *Growing through play* (pp. 36–46). Belmont, CA: Wadsworth.
Niemark, E. (1975). Longitudinal development of formal operational thought. *Genetic Psychology Monographs, 91*(2), 171–225.
Nolan, D. (1984, Fall). Children's feelings toward the latchkey child care arrangement and alternative programs: A review of the literature. *School Social Work Journal, 9*(1), 30–47.
Nyquist, L., Slivkin, K., Spence, J., & Helmrich, R. (1985, January). Household responsibilities in middle-class couples. *Sex Roles, 12*(1,2), 15–34.
Ornstein, A., & Levine, D. (1982). Sex, schools, and socialization. *The Educational Forum, 46*(3), 337–341.
Ouchi, W. G. (1981). *Theory Z.* Reading, MA: Addison-Wesley Publishing Company.
Packard, V. (1983). *Our endangered children: Growing up to a changing world.* Boston: Little Brown & Co., Inc.

Parnham, D. (1985, May 14). Living the good life. *U.S.A. Today*, p. 1.
Peale, N. V. (1982). *Dynamic imaging*. Old Tappan, NJ: Fleming H. Revell Company.
Pearson, D., & Shaw, S. (1982). *Life extension: A practical scientific approach*. New York: Warner Books, Inc.
Peck, E., & Granzig, W. (1978). *The parent test: How to measure and develop your talent for parenthood*. New York: Putnam.
Peck, E., & Leiberman, J. (1981) *Sex and birth control: A guide for the young*. New York: Schocken Books.
Pendergrass, R. (1985, March). Homework: Is it really a basic? *The Clearing House, 58*(7), 310–314.
Peterson, D. (1983). *Facilitating education for older learners*. San Francisco: Jossey-Bass.
Piaget, J. (1926). *The language and thought of the child*. New York: Basic Books.
Piaget, J. (1954). *The construction of reality in the child*. New York: Basic Books.
Piaget, J. (1970). Piaget's theory. In P. H. Mussen (Ed.), *Carmichael's manual of child psychology* (vol. 1). New York: John Wiley & Sons.
Piaget, J. (1981). *Intelligence and affectivity: Their relationship during child development*. Palo Alto, CA: Annual Reviews.
Pifer, A. (1983). Hard-headed arguments for investment in children. *High Scope Resources, 2*(2), 2–5.
Pitts, F. N. (1969). The biochemistry of anxiety. *Scientific American, 220,* 69–75.
Rank, O. (1973). *The trauma of birth*. New York: Harper & Row.
Reese, H. (1983). *Advances in child development*. New York: Academic Press.
Restak, R. (1985). *The brain*. New York: Bantam Books.
Roberts, D. F., & Backen, C. M. (1981). Mass communication effects. *Annual Review of Psychology, 32,* 307–356.
Roberts, T. (1975). *Four psychologies applied to education*. New York: John Wiley & Sons.
Rogers, C. (1983). *Freedom to learn*. Columbus: Charles E. Merrill.
Rook, K. S. (1984). Research on social support, loneliness and social isolation. In P. Shaver (Ed.), *Review of personality and social psychology*, vol. 5, (pp. 239–264). Beverly Hills, CA: Sage Publications.
Roseman, I. J. (1984). Cognitive determinants of emotion. In P. Shaver (Ed.), *Review of personality and social psychology*, vol. 5 (pp. 11–36). Beverly Hills, CA: Sage Publications.
Rosenthal, R., & Jacobson, L. (1968). *Pygmalion in the classroom*. New York: Holt, Rinehart & Winston.
Rosow, I. (1967). *Social integration of the aged*. New York: Macmillan.
Rutter, M. (1984, March). Resilient children. *Psychology Today, 18*(3), 56–65.
Santrock, J. W., & Yussen, S. R. (1984). *Children and adolescents*. Dubuque, IA:

William C. Brown Publishers.
Scarf, M. (1980, September). Images that heal. *Psychology Today, 14*(4), 32–46.
Schecter, H. (1975). The transcendental meditation program in the classroom. In D. W. Johnson & J. Farrow (Eds.), *Scientific research on the transcendental program.* New York: MIU Press.
Schuster, C., & Ashburn, S. (1986). *The process of human development.* Boston: Little Brown & Co.
Schwartz, H. (1984, February 17). Narrowing the black-white health gap. *The Wall Street Journal,* p. 24.
Segal, D. (1948). *Intellectual abilities in the adolescent period.* Bulletin No. 6. Washington, DC: Federal Security Agency, Office of Education.
Selye, H. (1956). *The stress of life.* New York: McGraw-Hill Book Company.
Shoup, B. (1980, May). Helping others yield personal growth for adolescents in Kansas City, Kansas. *Phi Delta Kappan, 61*(9), 633–635.
Silden, I. (1982, April-May). When adult children return home. *Modern Maturity, 25*(2), pp. 89-94.
Simon, S., Howe, L., & Kirschenbaum, H. (1978). *Values clarification.* New York: Hart Publishing Company.
Singer, J. (1973). *The child's world of make-believe.* New York: Academic Press.
Singer, J., & Singer, D. (1983). Psychologists look at television. *American Psychologist, 31*(7), 826–834.
Skinner, B. (1954). The science of learning and the art of teaching. *Harvard Educational Review, 24*(2), 86–97.
Skinner, B. (1971). *Beyond freedom and Dignity.* New York: Alfred A. Knopf.
Skinner, B. (1983, September). Origins of a behaviorist. *Psychology Today, 17*(9), 22–33.
Snarey, J. (1985). Cross-cultural universality of social-moral development: A critical review of Kohlbergian research. *Psychological Bulletin, 97*(2), 202–232.
Snygg, D., & Combs, A. (1949). *Individual behavior.* New York: Harper & Row.
Spainer, G., & Hanson, S. (1982). The role of extended kin in the adjustment to marital separation. *Journal of Divorce, 5*(1), 33–48.
Sparling, J., & Lewis, I. (1979). *Learning games for the first three years.* New York: Walker.
Sperry, R. (1982). Some effects of disconnecting the cerebral hemispheres. *Science,* 217, 1223–1226.
Spivak, E. (1985, Summer). The joy of home birth. *Medical Selfcare.* pp. 35–57.
Sroufe, L. A. (1982). Attachment and the roots of competence. In H. Fitzgerald & T. Carr (Eds.), *Human development,* (pp. 94–98). Guilford, CT: Dushkin Publishing Group.

Stamstead, M. (1985, Summer). The Intergenerational Educational Volunteer Network Act of 1985. *Intergenerational Clearinghouse, 3*(1), 2.

Stephen, T. (1985, November 20). *Exercise and adult mental health.* Paper presented at the American Public Health Association Annual Meeting in Washington, DC.

Sternberg, R. (1986). *Intelligence applied.* New York: Harcourt Brace Jovanovich.

Stewart, D. (1985, August). Teachers aim at turning loose the mind's eye. *Smithsonian, 16*(5), 44–54.

Stieglitz, E. (1952). *The second forty years.* Philadelphia: J. B. Lippincott.

Strom, R. D. (1975). Education for a leisure society. *The Futurist, 9*(2), 93–97.

Strom, R. D. (1981). *Growing through play.* Belmont: Wadsworth.

Strom, R. D. (1983). Expectations for educating the gifted and talented. *The Educational Forum, 47*(3), 279–303.

Strom, R. D, & Strom, S. (1985). *Becoming a better grandparent.* In K Struntz & S. Reville (Eds.), *Growing together: An intergenerational sourcebook* (pp. 57–60). Washington, DC: American Association of Retired Persons & The Elvirita Lewis Foundation.

Strom, R. D., & Strom, S. (1986). Family communication and television. *Television and Families, 9*(2), 29–33.

Super, D. E., & Hall, D. T. (1978). Career development: Exploration and planning. *Annual Review of Psychology, 20,* 333–372.

Swift, W. (1985, July). Assessment of the bulimic patient. *American Journal of Orthopsychiatry, 55*(3), 384–396.

Taylor, C. (1968). Cultivating new talents: A way to reach the educationally deprived. *Journal of Creative Behavior, 2*(2), 83–90.

Teri, L., & Lewinson, P. (1985). *Geropsychological assessment and treatment.* New York: Springer Publishing Company.

Terman, L. W. (1925). *Mental and psychical traits of a thousand gifted children, (vol. 1).* Stanford: Stanford University Press.

Terman, L. W., & Merrill, M. A. (1937). *Measuring intelligence.* Boston: Houghton Mifflin Company.

Terman, L. W., & Merrill, M. A. (1960). *Stanford-Binet Intelligence Scale.* Boston: Houghton Mifflin Company.

Terman, L. W., & Oden, M. (1947). *The gifted child grows up.* Stanford: Stanford University Press.

Thorndike, E. L. (1927). *The measurement of intelligence.* New York: Teachers College, Columbia University.

Thorndike, R. L. (1948). The growth of intelligence during adolescence. *Journal of Genetic Psychology, 72,* 111–115.

Thorp, K. (1985). *Intergenerational programs.* Madison: Wisconsin Positive Youth Development Initiative, Inc.
Thurstone, L. L. (1938). Primary mental abilities. *Psychometric Monograph I.* Chicago: University of Chicago Press.
Thurstone, L. L. (1955). The differential growth of mental abilities. Bulletin No. 14. Chapel Hill, NC: University of North Carolina Psychometric Laboratory.
Tice, C. (1985). *The states speak: A report on intergenerational initiatives.* Ann Arbor, MI: New Age Inc.
Tizard, B., & Hughes, M. (1985). *Young children learning.* Cambridge: Harvard University Press.
Toffler, A. (1970). *Future shock.* New York: Random House.
Toffler, A. (1980). *The third wave.* New York: William Morrow & Co.
Tomlinson-Keasey, C. (1985). *Child development: Psychological, sociocultural and biological factors.* Homewood: The Dorsey Press.
Torrance, E. P. (1967). *Understanding the fourth grade slump in creative thinking.* Washington, DC: U.S. Office of Education.
Torrance, E. P. (1979). *The search for satori and creativity.* Buffalo, NY: Creative Education Foundation.
Torrance, E. P. (1984). *Mentor relationships.* Buffalo, NY: Bearly Limited.
Torrance, E. P. (1986). Glimpses of the promised land. *Roeper Review, 8*(4), 246–251.
Torrance, E. P. (1968a). Teaching gifted and creative learners. In M. Wittrock (Ed.), *Handbook of research on teaching.* New York: Macmillan.
Trotter, R. J. (1983, August). Baby face. *Psychology Today, 17*(3), 14–20.
Turner, R., & Aschner, L. (1985). *Evaluating behavior therapy outcome.* New York: Springer Publishing Co.
Wallach, M., & Wallach, L. (1985, February). How psychology sanctions the cult of self. *The Washington Monthly,* pp. 46–56.
Wallat, C., & Piazza, C. (1986). The nation responds: Directions for literacy. *Theory Into Practice, 25*(2) 141–147.
Wallerstein, J. (1984). Helping children of disrupted families. *Elementary School Guidance and Counseling, 19*(1), 19–29.
Wallerstein, J. S., & Kelly, J. B. (1980, January). California's children of divorce. *Psychology Today, 13*(8), pp. 67–76.
Wallis, C. (1983, June 6). Stress: Can we cope? *Time, 121*(23), 48–57.
Watson, J. B. (1919). *Psychology from the standpoint of a behaviorist.* Philadelphia: J. B. Lipincott.
Watson, J. B. (1930). *Behaviorism.* New York: W. W. Norton Company.
Wechsler, D. (1944). *Measurement of adult intelligence.* Baltimore: Williams & Wilkins.

Whalen, R. E., & Simon, N. G. (1984). Biological motivation. *Annual Review of Psychology, 35,* 257–276.

White, B. (1985). *The first three years of life.* New York: Prentice-Hall Press.

Whitley, E., & Duncan, R. (1985). Life-long learners together. In K. Struntz & S. Reville (Eds.), *Growing together: An intergeneratinal sourcebook,* (p. 53–54). Washington, DC: American Association of Retired Persons and The Elvirita Lewis Foundation.

Williams, T. M., & Kornblum, W. (1985). *Growing up poor.* Lexington, MA: D.C. Heath & Company.

Wilson, K. (1983, July). Vision and hearing. *Senior World, 5*(5), pp. 1, 2, 10.

Winick, M. (1985, October). Search for the perfect pregnancy. *American Health, 4*(9), 58–61.

Wittrock, M. (1985). Education and recent neuro-psychological and cognitive research. In D. Benton and E. Zaidel (Eds.), *The dual brain.* Los Angeles: University of California at Los Angeles.

Wolf, M. (1985, February). The experience of older learners in adult education. *Lifelong Learning, 88*(5), 8–11.

Yankelovich, D. (1977). *Raising children in a changing society.* Minneapolis: General Mills.

Yankelovich, D. (1979). *Family health in an era of stress.* Minneapolis: General Mills.

Yankelovich, D. (1982). *New rules.* New York: Random House, Inc.

Yates, M. R., Saunders, R., & Watkins, J. F. (1980, June). A program based on Maslow's hierarchy helps students. *Phi Delta Kappan, 61*(10), 712–3.

NAME INDEX

Achenback, T., 173
Ageton, S., 123
Aiken, L., 76
Allcorn, S., 160
Allman, W., 59, 61, 67, 123, 125
Alzheimer, A., 106
American Medical Association, 156
Anderson, L., 116
Anderson, R., 211
Apgar, V., 52, 54, 57
Ardrey, R., 234, 250, 251
Arlin, P., 28
Aschner, L., 63, 69
Ashburn, S., 66
Averill, J., 112, 129

Backen, C., 253
Baldwin, C., 35, 101
Baltes, P., 107
Bandura, A., 29, 36

Bardwick, J., 225
Bayley, N., 98
Beck, J., 52, 54
Benedict, R., 38, 217
Bengston, V., 198
Bennett, W., 54, 65, 78
Benson, H., 51
Berger, R., 73
Berk, S., 220
Bezold, C., 202
Binet, A., 80
Birren, J., 39
Bonner, H., 28
Boston, B., 100
Bower, T., 112, 146, 231, 232
Bowlby, J., 113, 205, 232
Bradley, R., 128
Bradshaw, J., 82
Brazelton, T., 33, 114, 174, 231
Brent, D., 67
Brewer, G., 53, 176, 233

Brewer, T., 53, 176, 233
Bronfenbrenner, U., 34, 233
Brubaker, T., 42, 225, 255
Bruner, J., 91, 92, 130, 262
Buhler, C., 28
Bumpass, L., 210
Burish, T., 72
Butler, R., 42, 73, 108, 137, 195

Cagle, M., 102
Cannon, W., 110, 111
Caplan, F., 84, 147
Carey, J., 284
Carnegie Corporation, 34, 158, 162, 163
Cates, W., 54
Cezanne, P., 104
Chandler, T., 181
Chance, P., 144
Children's Defense Fund, 207
Chomsky, N., 267
Cloward, R., 152
Cohen, P., 152
Colangelo, N., 35, 101
Combs, A., 26
Comfort, A., 221–222
Cook, B., 181
Corey, L., 54, 72
Cornell, C., 205
Cousins, N., 51
Cox, J., 100
Crane, H., 105

Daniel, N., 100
Darnell, J., 32
Darwin, C., 104
DeBono, E., 145, 165
Dentzer, S., 226, 256
Dillon, J., 278, 279
Donaldson, M., 94, 272
Donne, J., 229
Donovan, S., 213

Driscoll, F., 213
Dugovics, D., 181
Duncan, R., 198, 287
Dunn, R., 90, 91
Durkheim, E., 256

Edelbrock, C., 173
Eichhorn, D., 213
Eisenberg, A., 75, 136, 192
Eisenberg, H., 75, 136, 192
Elderhostel, 167, 168
Elkin, F., 181, 239
Elliott, D., 123
Ennis, T., 106
Epstein, H., 86–88
Erikson, E., 27, 36, 46, 47, 113–115, 117, 125, 129, 133, 137, 189, 190, 251, 264
Eron, L., 128, 182, 209
Eurich, N., 161
Euster, G., 168

Family Service Association of America, 280
Farran, D., 237
Feldman, H., 131, 210
Feldman, M., 131, 210
Felsenfeld, G., 22, 32
Fogg, R., 165
Fowers, B., 282
Fraiberg, S., 113, 206, 232
Freud, S., 18, 23, 26, 27, 36, 38, 56, 125, 126, 172, 188
Friedrich, O., 33, 147, 231
Fromm, E., 262
Frymier, J., 118

Gagne, E., 36, 266
Gagne, R., 30, 31
Gale, A., 42, 76, 156
Galloway, C., 175

Garfinkel, P., 70
Gardner, H., 80, 92
Garner, D., 70
Gaylin, W., 110, 116, 165
Gelles, R., 205
Gelman, D., 255
Gerler, E., 211
Ghiselin, B., 104
Glass, B., 22, 23
Gleason, H., 42, 73, 108, 137
Glieberman, H., 189
Gober, P., 226, 285
Goldman, J., 36
Goldman, R., 36
Goleman, D., 51, 122, 126, 201, 212
Goodstein, H., 246
Granzig, W., 130
Green, R., 131
Guilford, J., 80, 98, 104
Gurin, J., 62, 68, 78

Hagestad, G., 230
Hall, D., 183
Hall, G.S., 18, 38
Halverson, C., 62
Handel, G., 181, 239
Hanson, S., 227
Harper, P., 23, 32, 53
Harrington, B., 100
Harrington, J., 100
Harris, G., 62, 68, 78
Harris, I., 277
Hasselbring, T., 246
Havighurst, R., 46
Hawley, W., 246
Helmrich, R., 221
Helms, D., 30
Hemingway, E., 105
Hillson, H., 158
Hofmeister, A., 260
Holmes, K., 54, 72
Howard, I., 277
Howard, T., 23

Howe, L., 242
Howell, M., 60
Huessman, L., 209
Hughes, M., 237
Huizinga, D., 123
Hunt, S., 220

Isakson, S., 102
Izard, C., 66, 111

Jacobson, L., 178
James, W., 18, 23, 110, 112
Jenkins, C., 78, 111, 219
Jensen, A., 22, 87
Jersild, A., 47
Juhasz, A., 37
Jung, C., 23

Kastenbaum, R., 138, 139
Katz, L., 267
Kelly, G., 75
Kelly, J., 211
Kennell, J., 206, 233
Kerman, S., 178
Kilmann, R., 189
Kim, J., 58, 72, 85
Kinney, J., 211
Kirschenbaum, H., 242
Klaus, M., 33, 206, 233
Kobasa, S., 201
Koffka, K., 25
Kohlberg, L., 46, 47, 248, 249
Kohler, W., 25
Kolata, G., 65
Kolevzon, M., 131
Kornblum, W., 58
Kornhaber, A., 198, 227
Kravitz, D., 25, 280, 283
Krech, D., 266
Kubler-Ross, E., 139
Kulik, C., 152

Lamaze, F., 55
Lauer, J., 41, 133, 135
Lauer, R., 41, 133, 135
LeBow, M., 63
Lefkowitz, M., 209
Lehman, H., 101
Leiberman, J., 132
Leiblum, S., 75
Leich, J., 167
Levine, D., 220
Levinson, D., 39, 222
Levy, S., 72
Lewin, K., 25
Lewinson, P., 42, 106
Lewis, I., 147, 266
Lewis, M., 195
Linden, F., 226
Lindsley, D., 111
Lippitt, P., 152
Lipsitz, J., 39, 216
Long, L., 243, 244
Long, T., 243, 244
Lowes, J., 104
Lucretius, 79
Lyon, H., 185
Lytle, V., 143

Maccoby, E., 230
Maddock, J., 189, 194
Malatesta, C., 66, 111
Malecki, M., 138
Maslow, A., 28, 47, 171, 172, 197, 203, 262, 265
May, R., 262
Mayer, J., 63
Mayerowitz, B., 72
Mayman, M., 28
McCabe, E., 151
McDougall, W., 18, 23
McGaugh, J., 58
McGrath, J., 25, 280, 283
McMillan, P., 207
Mead, M., 38, 217

Meddis, S., 59
Meir, J., 60
Menninger, K., 28
Merrill, M., 97
Miller, M., 256
Mobilization for Youth Project, 152
Moos, R., 222
Murphy, M., 213
Mussen, P., 37
Myers, F., 158

Naisbitt, J., 23, 39, 145, 261
National Center for Research in Vocational Education, 183
National Institute on Aging, 108
Nettleton, N., 82
Neubeck, G., 189, 194
Neugarten, B., 40, 195, 224, 230
Nichols, M., 193
Nicholson, J., 69, 193, 224
Nicolayson, M., 234
Niemark, E., 94
Nolan, D., 243, 245
Nyquist, L., 221

Oden, M., 47
Olson, D., 282
Ornstein, A., 220
Ouchi, W., 190

Packard, V., 134, 211, 243
Parents Without Partners, 281
Parnes, S., 102
Parnham, D., 188
Peale, N.V., 196
Pearson, D., 196
Peck, E., 149, 130, 132
Pendergrass, R., 154
Personnel and Guidance Association of North Carolina, 211
Pervin, L., 75

Pestalozzi, J., 18
Peterson, D., 168
Peterson, R., 128, 182
Piaget, J., 18, 46, 83–84, 86, 88–89, 93–94
Piazza, C., 273
Pieper, H., 167
Pifer, A., 88
Pittman, D., 35, 101
Pitts, F., 202
Proctor, W., 51
Pruyser, P., 28

Quinn, T., 54, 72

Rank, O., 56
Reese, H., 63
Restak, R., 82, 87, 96
Rifkin, J., 23
Roberts, D., 253
Roberts, T., 26
Robertson, J., 198
Rogers, C., 28, 261, 262
Rook, K., 136, 281, 285
Roseman, I., 129
Rosenholtz, S., 246
Rosenthal, R., 178
Rosow, I., 257
Rutter, M., 120

Santrock, J., 125
Saunders, R., 178
Scarf, M., 196
Schaie, K., 108
Schecter, H., 213
Schuster, C., 66
Schwartz, H., 76, 84
Segal, D., 94
Selye, H., 200
Shaw, G., 178
Shaw, S., 196

Sheehy, G., 262
Shoup, B., 184
Silden, I., 192, 218
Simon, N., 58, 171
Simon, S., 242
Simon, T., 80
Singer, D., 268
Singer, J., 148, 268
Skinner, B.F., 24, 171
Slivkin, K., 221
Snarey, J., 249
Snygg, D., 26
Spanier, G., 227
Sparling, J., 147, 266
Spence, J., 221
Sperry, R., 82
Spivak, E., 56
Sroufe, L.A., 173
Stamstead, M., 146
Stein, G., 105
Stephen, T., 69
Sternberg, R., 81
Stewart, D., 148
Stieglitz, E., 40
Strom, R., 150, 159, 169, 185, 258, 270
Strom, S., 169, 258, 270
Super, D., 183
Sussman, M., 189, 194
Swift, W., 71

Taylor, C., 81
Teeter, T., 128
Teri, L., 42, 106
Terman, L.M., 20, 47, 80, 96, 97
Thorndike, E., 98
Thorndike, R., 98
Thorp, K., 42, 60, 143
Thurstone, L., 80, 98
Tice, C., 143, 197–198
Tizard, B., 237
Toffler, A., 39, 143–144
Tomlinson-Keasey, C., 23

Torrance, E.P., 39, 99, 144, 151
Trotter, R., 175, 231, 266
Turner, J., 30
Turner, R., 63, 69

Victor, J., 62

Waldor, L., 209
Wallach, L., 191, 251, 263
Wallach, M., 191, 251, 263
Wallat, C., 273
Wallerstein, J., 211
Wallis, C., 200
Watkins, J., 178–179
Watson, J.B., 23
Wechsler, D., 80

Weinberg, N., 58
Whalen, R., 58, 171
White, B., 33, 84, 114, 147, 174, 205, 231, 266
Whitley, E., 198, 286, 287
Williams, T., 58
Wilson, K., 74
Winick, M., 53, 55
Wittrock, M., 83, 92, 96
Wolf, F., 181
Wolf, M., 161

Yankelovich, D., 200, 262, 271
Yates, M., 178–179
Yussen, S., 125

Zonn, L., 226, 285

SUBJECT INDEX

Adolescence
  accidents and suicide, 66–68, 125
  attitudinal crisis, 124–125
  barriers to adulthood, 38–39
  brain reorganization, 95–96
  defense mechanisms, 125–127
  definition of, 37–39, 217
  drug abuse, 67–68
  family communication, 275–279
  formal thinking, 93–94
  menarche, 65
  moral decisionmaking, 248–249
  motivational needs, 187–188
  physical change, 38, 65–66
  roles, 274–277
  stress sources, 214–217
Ancestry, 31–32
Attachment, 33, 113, 139, 173–174, 205–206, 232–233

Behaviorism
  origins of, 23, 24
  modification, 24
  operant conditioning, 24
  reinforcement, 174–175
Birth, 55–56
Boredom, 183–184, 197, 217–218, 250–251
Brain
  Alzheimer's disease, 106–107
  emotions, 110–112
  growth, 84, 86–87, 90, 93, 97
  hemispheres, 82–83, 92, 99
  malnutrition, 58, 84–85
  motivational drives, 173
  processes, 90–92, 99
  reorganization, 95–96

Cancer
  diet guidelines, 71–72
Climacteric, 75, 194
Communication
  family televiewing, 268–270

feelings, sharing, 120–123, 263–265
group processes, 123–124, 197, 215–216, 272–273
intergenerational, 139, 228, 286–287
levels of discourse, 263–265
listening skills, 270–271, 282–283
marital success, 133–135, 188–189, 280–282
parent guidelines, 180–181, 275–279
retirement peers, 285–286
school conferences, 271–272
speech development, 44, 266–268
therapy groups, 192, 197, 211, 226, 261–262, 279–280, 283–284
Conflict resolution, 163–166
Creativity
assessment of, 99–100
brain hemispheres, 92, 99
imaging, 196
influence of solitude, 148–151
job success, 101–102
motivation for, 151
problem finding, 28
thinking process, 102–105

Data gathering techniques
biographies, 18
clinical, 20
cross sectional, 19
experimental, 20
longitudinal, 19
observations, 18, 30, 33
questionnaire, 18
Defense mechanisms, 125–127
Diet
anorexia, 70–71
brain, 58
bulimia, 70–71
cancer, 71–72
hunger, 57–58
obesity, 62–63, 69–71, 173
Diseases
Alzheimer's, 106–107
cancer, 71–72
childhood, 59, 61–63
genetic, 53
sexually transmitted, 53–54, 72–73
Divorce, 210–211, 281–282
Dropout, school
ethnic rate, 158
gifted students, 185
illiteracy, 273
reduction, 152–154, 157–159
Drugs
accidents, 67–68, 123
designer, 123
emotional change, 123–124
group discussion, 68, 123–124
physiology, 202
prenatal, 52–54
suicide, 67
therapy, 111, 123–124

Early adulthood
attitudinal crisis, 129–130
conflict resolution, 163–166
definitions of success, 39, 40, 188–190
eating disorders, 70–71
higher education, 159–163
identity, 250–252, 264–265
marital success, 41, 133–135, 280–282
occupational burnout, 220–221
parenting satisfactions, 130–132
physical fitness, 69
therapy groups, 279–280
work ethic, 189–190, 251
Early childhood
accidents, 59, 61
attitudinal crisis, 115–117
brain growth, 86–88
creative behavior, 148–151

language, 44, 89
peer relations, 208–209, 233–237
preoperational thinking, 88–89
schooling, 34–35, 88, 233–237
televiewing, 268–270
value acquisition, 175–177
Emotion
  aggression, 116–117, 206–207
  attachment, 112–115, 173–174
  attitudinal crises, 113–118,
    124–125, 129–130, 132–133,
    137
  defense mechanisms, 125–127
  grief, 138–139, 225–226,
    286–287
  theories of, 110–112, 128–129
  withdrawal, 67, 123–124, 202,
    206–207

Genetics
  ancestry, 31–32
  counseling, 32, 53
  diseases, 53, 107
  engineering, 31, 32
  genius, 19
  lifespan, 76–77
  motivation, 171
  theory, 42
  traits, 21, 22
Gestalt, 24–26
Giftedness
  dropout rate, 185
  genius, 19
  identification, 98–100
  motivation, 185
  schooling, 100, 185
Grandparents
  curriculum for, 257–258
  educational needs, 168–169
  frail health, 77–78, 197–198,
    253–255
  guidance source, 183
  intergenerational programs,
    145–146, 197–198

  role shift, 197–198, 227, 257–258

Handicaps
  acceptance of, 238–239
  blind, 185–186
  definition of, 239
  educational law, 239–240
  mainstreaming, 240
  motivation, 185–186
  retardation, 240–241
Health risks
  accidents, 59, 66–68
  cancer, 71–72
  child abuse, 60–61, 203–205
  disease, 59–60, 73
  drug abuse, 52–54, 67–68, 72,
    123–124
  eating disorders, 62–63, 69–71,
    73–74, 243
  elderly abuse, 255
  exercise, lack of, 61–62, 68–69
  malnutrition, 57–58, 84–85, 203
  smoking, 54, 68, 72
  stress, 200–201, 219
  suicide, 67–68, 125, 256–257
Hearing
  childhood, 65
  infancy, 231
  midlife, 74
  old age, 78
Homework practices, 154–156
Humanistic psychology
  hierarcy of needs, 172
  phenomenology, 26
  proactive behavior, 29
  self actualization, 28, 262, 265

Information processing, 30–31
Infancy
  abuse, 51–61, 203–206, 233
  attachment, 33, 113, 173–174,
    205–206, 232–233

attitudinal crisis, 114–115, 232–233
growth tests, 57
instruction, 147–148
nutrition, 84–85
premature, 84–85
sensorimotor thinking, 86, 146–147
social awareness, 231–232
speech, 266–268
Intelligence
aging, 105–108
aptitudes, 94–95
assessment, 96–97, 148
definitions, 80–81
exceptionality, 98–100, 185–186
growth, 34, 96–98
heredity, 21–22, 87
Intergenerational relationships
Child Care Act, 59, 60
conflicts, 145–146
Family Friends project, 60
Foster Grandparents, 60
mutual needs, 78, 145–146, 196–198, 258
sharing perspectives, 183, 228
suicide prevention, 67
successful contacts, 145–146, 286–287
teaching potential, 145–146, 156, 196–198

Later adulthood
abuse, 255
aging theories, 41–43, 76–77
attitudinal crisis, 137–139
custodial care, 77, 78, 138–139, 253–255
grandparenting, 168–169, 197–198, 257–258
intellectual performance, 107–108
life expectancy, 76–77
physical changes, 77–78

retirement, 226–227, 255–257, 285–286
school access, 166–168
self assessment, 196
senility, 77, 105–107
stress, 225–227
Later childhood
attitudinal crisis, 118–120
concrete thinking, 89–90
drug abuse, 123–124
fears and anxieties, 121–123
goalsetting motivation, 182–183
latchkey experience, 243–245, 269
peer teaching, 152–154
physical changes, 36–37, 63–64
stress reduction, 212–213
vision and hearing, 64–65
Loneliness, 136–137, 196–197, 225–226, 256–257
divorce, 210–211, 281–282
empty nest, 191–192

Mainstreaming, 239–240
Menopause, 74–75, 193–194
Middle age
attitudinal crisis, 132–133
caring for parents, 253–255
corporate classroom, 161–163
lifestyle choices, 40–41
loneliness, 136, 191–192
marriages, 41, 133–135
physical fitness, 73–74
self evaluation, 222–224, 262–263
sexual behavior, 74–75
vision and hearing, 74
Middle childhood
attitudinal crisis, 118–120
concrete thinking, 89–90
divorce effects, 209–211
fitness and growth, 61–62
group processes, 272–273
parental pressure, 180–181
peer abuse, 208–209

peer acceptance, 180, 240–241
reading, 273–274
school conferences, 271
Motivation
adolescent needs, 187–188
boredom, 183–184, 197, 217–218
clinical studies, 20–21
failure, 152–154, 157–159, 177, 180–181, 185, 273
fantasy play, 175–177
goalsetting, 181–183
hierarchy of needs, 172
meanings of, 171
parental pressure, 180–181
peer approval, 180
Pygmalion effect, 177–178
reinforcement, 174–175
self concept, 157–159, 262–263
success, 188–190

Neonate, 56, 57

Obesity
adult, 69, 73–74
cancer, 71
childhood, 62–63
health risks, 69–70
hypothalamus, 173
theories of, 69–70

Parenting
child abuse, 60–61, 203–206, 233
communication, 181–183, 276–279
divorce, 209–211
education, 163–166
family conflict, 163–166, 180–181
marital success, 41, 133–135, 280–282
role, 274–275
satisfactions, 130–132

televiewing guidelines, 268–270
values teaching, 175–177
Peers
accepting differences, 237–241
group approval, 180, 255–257, 285–286
group processes, 123–124, 197, 272–273, 285–286
intimidation, 208–209
teaching, 152–154
territoriality, 233–237
Play
acquiring values, 175–177
dominion, 89, 233–237
games and toys, 147–148, 175–177
preschool, 34–35
solitary, 148–151
Prenatal
adoption, 32
amniocentesis, 54–55
birth defects, 52–53
critical periods, 52–54, 84
malnutrition, 84–85
placenta, 54
ultrasound test, 55
Psychoanalysis
defense mechanisms, 125–127
origins, 26–27
personality structure, 26–28, 113–114, 125–126
psychosomatic illness, 200, 219
Psychological viewpoints
hereditarian, 21–23
behaviorism, 23–24
Gestalt-field, 24–26
psychoanalytic, 26–27
humanistic, 27–29, 172
social learning, 29–30, 230
information processing, 30–31
Puberty, 38, 65–66

Self concept
defense mechanisms, 125–127

identity, 193–195, 250–252, 264–265
imaging, 196
locus of control, 127–128
narcissism, 191
school curriculum, 157–159, 250–252
self evaluation, 222–224, 246–247, 250–252
success, 39–40, 188–190
Sexuality
climacteric, 75, 194
disease, 53–54, 72–73
marriage, 135, 188–190, 281–282
menarche, 65
menopause, 74–75, 193–195
Smoking, 54, 68, 72
Social competence
accepting others, 237–241
adult challenges, 249–250
attachment, 33, 113, 173–174, 205–206, 232–233
definitions of, 230
grandparent role, 197–198, 227, 257–258
group processes, 123–124, 197, 215–216, 272–273
identity, 193–195, 250–252, 264–265
marital success, 41, 133–135, 280–282
moral development, 241–243, 248–249
mutual rights, 235–236
parental satisfaction, 130–132
school discipline, 237, 245–247
Stress
abusive behavior, 59–61, 203–206, 208–209, 254–255

divorce, 209–212
drugs, 202
ethnicity, 207–208
independence, 249–250
latchkey, 243–245
meanings of, 200–201
reduction, 203, 212–217
retirement, 225–227
school influence, 154–159
symptoms, 206–207
Suicide
adolescent, 66–68, 125
causes of, 256–257
elderly, 256–257

Television
family guidance, 253, 268–270
higher education, 159–163
Thinking
analytic, 143
creative, 102–105, 144–145
concrete, 89–90
critical, 144
formal, 91–92
imaging, 196
learning styles, 90–92, 99, 118–119
preoperational, 88–89
problem finding, 45
sensorimotor, 86, 146–147

Vision
astigmatism, 273–274
childhood, 64–65, 273
infancy, 231
midlife, 74
later life, 77–78